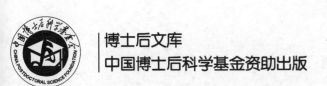

博士后文库
中国博士后科学基金资助出版

金属有机框架材料在药物递送领域的应用

曹 健 著

科 学 出 版 社
北 京

内 容 简 介

金属有机框架（MOFs）材料在药物递送领域的应用前景广阔，需要不断深入研究提升其性能，从而推进其在实际治疗中的应用。本书介绍了MOFs 材料在药物递送方面的独特优势和最新研究进展。全书分七章，第1 章概述基于 MOFs 药物递送应用；第 2 章详细介绍金属有机框架材料的制备；第 3 章针对 MOFs 在药物递送领域的应用介绍载体设计方法；第 4章介绍靶向与修饰策略；第 5 章介绍 MOFs 在抗癌领域的应用；第 6 章介绍 MOFs 在抗菌、抗炎、生物成像等其他领域的应用；第 7 章介绍 MOFs在药物递送领域应用未来发展与挑战。

本书可供药剂、材料、生命等相关领域科研工作者查阅参考。希望能为致力于研究新型药物递送方法的广大同行提供借鉴，同时为广大致力于纳米载药研究的科研工作者推动 MOFs 从实验室研究走向临床应用提供参考。

图书在版编目（CIP）数据

金属有机框架材料在药物递送领域的应用 / 曹健著. -- 北京：科学出版社，2025. 3. --（博士后文库）. -- ISBN 978-7-03-081546-0

Ⅰ. TQ460.1

中国国家版本馆 CIP 数据核字第 2025BD6630 号

责任编辑：刘　冉 / 责任校对：杜子昂

责任印制：徐晓晨 / 封面设计：蓝正设计

科学出版社 出版

北京东黄城根北街 16 号

邮政编码：100717

http://www.sciencep.com

北京厚诚则铭印刷科技有限公司印刷

科学出版社发行　各地新华书店经销

*

2025 年 3 月第　一　版　开本：720×1000　1/16

2025 年 3 月第一次印刷　印张：13

字数：260 000

定价：108.00 元

（如有印装质量问题，我社负责调换）

"博士后文库"序言

1985年，在李政道先生的倡议和邓小平同志的亲自关怀下，我国建立了博士后制度，同时设立了博士后科学基金。30多年来，在党和国家的高度重视下，在社会各方面的关心和支持下，博士后制度为我国培养了一大批青年高层次创新人才。在这一过程中，博士后科学基金发挥了不可替代的独特作用。

博士后科学基金是中国特色博士后制度的重要组成部分，专门用于资助博士后研究人员开展创新探索。博士后科学基金的资助，对正处于独立科研生涯起步阶段的博士后研究人员来说，适逢其时，有利于培养他们独立的科研人格、在选题方面的竞争意识以及负责的精神，是他们独立从事科研工作的"第一桶金"。尽管博士后科学基金资助金额不大，但对博士后青年创新人才的培养和激励作用不可估量。四两拨千斤，博士后科学基金有效地推动了博士后研究人员迅速成长为高水平的研究人才，"小基金发挥了大作用"。

在博士后科学基金的资助下，博士后研究人员的优秀学术成果不断涌现。2013年，为提高博士后科学基金的资助效益，中国博士后科学基金会联合科学出版社开展了博士后优秀学术专著出版资助工作，通过专家评审遴选出优秀的博士后学术著作，收入"博士后文库"，由博士后科学基金资助、科学出版社出版。我们希望，借此打造专属于博士后学术创新的旗舰图书品牌，激励博士后研究人员潜心科研，扎实治学，提升博士后优秀学术成果的社会影响力。

2015年，国务院办公厅印发了《关于改革完善博士后制度的意见》（国办发〔2015〕87号），将"实施自然科学、人文社会科学优秀博士后论著出版支持计划"作为"十三五"期间博士后工作的重要内容和提升博士后研究人员培养质量的重要手段，这更加凸显了出版资助工作的意义。我相信，我们提供的这个出版资助平台将对博士后研究人员激发创新智慧、凝聚创新力量发挥独特的作用，促使博士后研究人员的创新成果更好地服务于创新驱动发展战略和创新型国家的建设。

祝愿广大博士后研究人员在博士后科学基金的资助下早日成长为栋梁之才，为实现中华民族伟大复兴的中国梦做出更大的贡献。

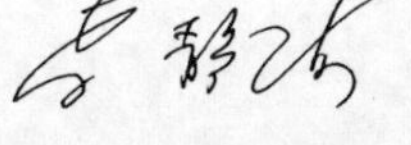

中国博士后科学基金会理事长

前　言

开发可控的药物递送系统对于减少副作用和增强药物的疗效至关重要。随着科技的快速发展，新型多孔材料——金属有机框架（metal organic frameworks，MOFs）材料逐渐进入人们的视野，其在药物递送领域中的应用成为当前研究热点之一。本书将介绍金属有机框架材料在药物递送方面的独特优势和最新研究进展，为科研工作者提供有关MOFs材料在药物递送领域的研究思路和方法。

MOFs 由有机配体和与其配位的金属离子或团簇组成，形成一维、二维或三维有序结构，利用丰富的阳离子和多种有机配体可以制备出具有不同拓扑结构、孔隙结构和尺寸的 MOFs。因其具有独特的物理化学性质，关于 MOFs 的研究已遍布气体储存、催化、分子分离、光电器件、药物缓释等诸多领域。

在药物递送领域中，MOFs 材料具有的多孔结构和大比表面积，使其有很高的载药能力。而且 MOFs 材料的孔径和结构可以通过调节金属离子和有机配体的组合来进行调控，以实现药物的精确控制释放，使 MOFs 材料在药物递送领域具有广泛的应用前景。另一方面，金属有机框架材料的化学稳定性良好，在酸碱环境和氧化还原条件下均具有较强的适应性，这使得 MOFs 材料在药物递送过程中能够保持稳定的结构，来确保药物的稳定释放。此外 MOFs 材料还具有良好的生物相容性，能够降低药物对人体的毒副作用，提高药物疗效。

目前，金属有机框架材料在药物递送领域的应用研究已经取得了一定的进展。科研人员采用 MOFs 材料，成功实现了抗癌药物、抗生素、抗菌、基因治疗药物等的精确控制释放。同时 MOFs 材料在生物成像、靶向给药、缓控释递送系统等方面的应用，也将为疾病诊断和治疗提供更为有效的方法。采取跨学科研究，如在材料科学、化学、生物学和医学等领域的交叉合作，将有助于推动 MOFs 材料在药物递送领域取得更多的突破性进展。

MOFs 材料在药物递送领域的应用前景广阔，为了进一步推进其发展，科研人员还需解决一些关键问题。例如，提高 MOFs 材料的药物负载量和稳定性，这就需要更深入地研究 MOFs 材料的合成方法，开发具有更高负载量和更好稳定性的新型 MOFs 材料。还需研究 MOFs 材料与生物体的相互作用机制，了解 MOFs 材料在体内的分布、代谢和排泄情况，以及其对生物体的毒副作用，评估 MOFs 材料作为药物递送载体的安全性，这就需要通过动物实验和临床研究来获取相关数据，为 MOFs 材料在药物递送领域的应用提供科学依据。还需进一步调控其孔径和结构，实现药物的快速、持续和可控释放，实现绿色大规模制备，推进其商业化应用。

目前，MOFs 材料在药物递送领域的应用研究很多还处于实验室阶段，需要进一步推进其在实际治疗中的研究。不断深入研究、优化和完善 MOFs 材料的合成，提升其在药物负载、药物释放等方面的性能。推动 MOFs 从实验室研究走向临床应用，为病患带来更好的治疗选择。相信在不久的将来，MOFs 将成为药物递送领域的一种重要工具，为人类健康事业的发展做出更大的贡献。本书对当前金属有机框架材料在药物递送领域的应用进行了介绍，以帮助研究者了解 MOFs 在该领域的现状和挑战。本书编写过程中，得到了所在单位廊坊师范学院化学与材料科学学院的支持，在此向多位同仁的帮助表示感谢！希望本书能为该领域研究提供参考，为人类健康事业的发展贡献绵薄之力。

曹　健

2024 年 10 月

目　录

第1章 基于MOFs药物递送应用概论

1.1 什么是基于MOFs的药物递送

金属有机框架（metal organic frameworks，MOFs）材料是一类具有周期性网络结构的晶态多孔有机–无机杂化材料，它既不同于无机多孔材料，也不同于一般的有机配合物，而是一类兼具无机材料的刚性和有机材料的柔性双重特征的杂化材料。这些特征使其在现代材料研究方面呈现出巨大的发展潜力和诱人的发展前景。MOFs具有的多孔、大比表面积和多金属位点等诸多性能，使其在化学化工领域得到许多应用，例如气体储存[1]、分子分离[2]、催化[3]、药物缓释[4]、传感等。

金属有机框架材料的优势，包括其可控形态、可定制的直径、可调孔隙率、高比表面积、容易功能化和良好的物理化学性质等。研究者们已经设计和构建了各种基于MOFs的纳米平台，以满足药物递送平台治疗的各种需求。多孔结构使MOFs 成为不同药物递送的优秀候选材料：金属离子和有机配体选择的灵活性使得制备具有内在活性的MOFs和进一步设计MOFs-药物协同系统成为可能；多种其他类型的药物也可以使用MOFs作为前体或模板。我们见证了基于MOFs药物的巨大发展，其高负载能力和药物精确控制释放特性使得MOFs成为了新型药物递送系统的研究热点。

金属有机框架（MOFs）材料作为一类晶体材料，由有机配体和与其配位的金属离子或团簇组成，形成一维、二维或三维结构，是由含氧、氮等的多齿有机配体（大多是芳香多酸和多碱）与过渡金属离子自组装而成的配位聚合物，是具有周期性网络结构的晶态多孔材料[5]。自20世纪90年代中期以来，金属有机框架（MOFs）材料已经取得了显著进展。但是MOFs的合成虽然揭开了这一研究领域的序幕，但较低的孔隙率和化学稳定性限制了它的实际应用。为了克服这些缺点，研究者们开始探索和开发新型配位体，这些配位体包括阳离子、阴离子和中性的配位体，通过与金属离子形成稳定的配位聚合物，从而显著改善MOFs的性能。早期的MOFs主要由简单的配位化学合成，其孔隙率和化学稳定性都不高，这些初期的研究为后续的改进奠定了基础。研究者们认识到，通过选择合适的配位体和金属离子，可以显著提高 MOFs 的性能，特别是含羧基的有机阴离子配体和含氮杂环有机中性配体的引入，使得 MOFs 的孔隙率和化学稳定性得到了大幅提升。新的配位体不仅提高了 MOFs 的化学稳定性，还增加了其结构的多样性，使得 MOFs 在不同的应用场景中展现出独特的优势，如用于吸附分离 H_2[6]，作为催化剂[7]、磁性材料[8]和光学材料[9]等。此外，MOFs 作为一种超低密度多孔材料，它们可以通过选择金属节点和有机连接剂进行精确定制。这种易定制性使其能够在诸多领域进行应用，包括气体储存、分离、催化等。在药物递送（drug delivery）领域，它们的受控方式装载和释放治疗剂的能力与传统的药物输送系统相比，也具有显著优势（图 1.1）。

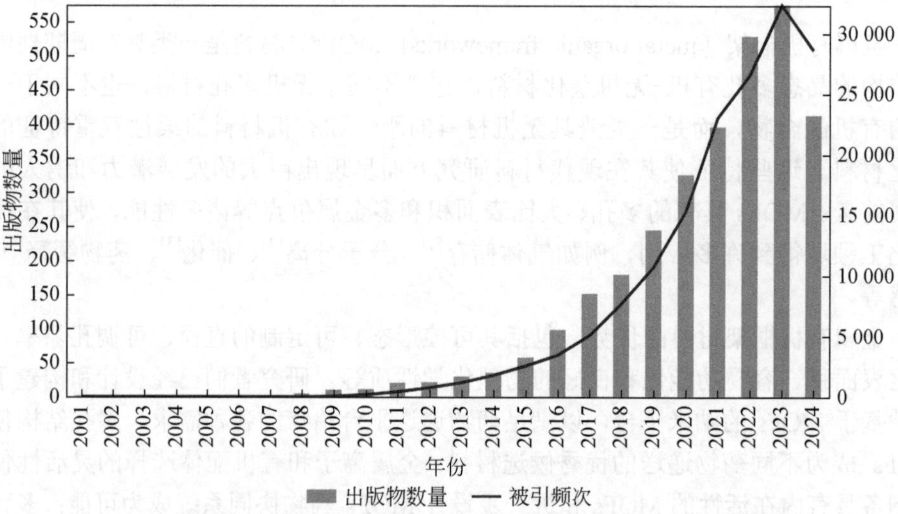

图 1.1　基于 MOFs & Drug delivery 关键词研究（来自 Web of Science）

一方面，MOFs 材料具有的多孔结构和大比表面积，使其具有很高的载药能力。而且研究人员可以通过调节金属离子和有机配体的组合来对 MOFs 材料的孔径和结构进行调控，实现药物的精确控制释放，从而使 MOFs 材料在药物递送领域具有广泛的应用前景。另一方面，金属有机框架材料的化学稳定性良好，对酸碱环境和氧化还原条件均具有较强的适应性，这保证 MOFs 材料在药物递送过程中能够保持稳定的结构，以确保药物的稳定释放。此外，MOFs 材料还具有良好的生物相容性，能够降低药物对人体的毒副作用，提高药物疗效。

MOFs 因其强大的药物吸收能力和可控的缓释过程，成为极具吸引力的药物传递载体，能够保护并将敏感药物分子精确传递到靶区。纳米级的 MOFs 可以吸附和释放各种药物/活性物质（如布洛芬[10]、姜黄素[11]、抗生素[12]、一氧化氮等）。MOFs 材料在装载其他药物后具有成像和治疗功能，可考虑用于多功能一体化靶向药物系统的研究。通过纳入其他客体材料，一些 MOFs 复合材料被开发出来，提供了额外功能，如外部触发药物释放、研究改进的药代动力学和诊断辅助。目前金属有机框架已被应用到许多纳米平台用于治疗恶性肿瘤［如光动力学治疗（PDT）[13]、化学动力学治疗[14]、化疗[15]］和改善细菌耐药性[16]等，从制药、疾病治疗到先进的药物给药系统中都有 MOFs 的应用研究。

随着研究不断深入，多种 MOFs 材料被用于药物递送领域。例如 MOF-5，这种材料以引入高孔隙率和高比表面积而闻名，有研究报道其可用于潜在的药物递送；HKUST-1（$Cu_3(BTC)_2$）可调控性强，将该 MOFs 功能化可以增强其药物负载和释放的功能；ZIF-8 的结构特征是其独特的四面体和八面体孔隙，这些孔隙相互连接形成一个高度有序且均匀的三维孔隙网络。其孔隙尺寸大约为 3.4 Å，而比表面积可以超过 1000 m^2/g，具有优异的吸附和分子筛选能力[17]。UiO-66 是一种耐高温的 MOFs 材料，以其卓越的稳定性与生物相容性而著称[18]，是生物医学应用中的主要候选者。这些关于 MOFs 载药的报道，表明 MOFs 在药物递送系统中有着广阔的应用前景。

在过去十年中，金属有机框架材料在药物递送中的发展取得了显著进展，MOFs 已经从递送小分子药物过渡到更复杂的应用，包括用于靶向治疗的刺激响应系统[19]，展示了它们在应对与传统药物递送系统相关的挑战方面的多功能性和潜力。这种演变反映了 MOFs 在创造更高效、更有针对性和生物相容性的药物递送解决方案方面日益增长的重要性。MOFs 因其高比表面积和可调控的孔结构的优点而具有高效装载药物的能力。通过调整 MOFs 的孔径和功能化表面，可以实现对药物释放速率的精确控制。例如，通过改变 MOFs 材料的 pH 敏感性组分，可以设计出在特定 pH 条件下触发药物释放的系统。某些 MOFs 可以用作光敏剂的载体，用于光动力学治疗。MOFs 的结构可以保护光敏剂免受自我猝灭，并使得激活光在体内更有效地转换为能够杀死癌细胞的活性氧。Zhao 等[20]构建了 DNA 功能化卟啉 MOF（porMOF）药物递送系统，porMOF 作为光敏剂和药物递送载体可以起到整合光动力学治疗（PDT）和化疗的作用。通过将 MOFs 与不同治疗方式（如化疗、光疗、超声治疗）相结合，可以构建多功能的药物递送系统，实现不同治疗方案的协同增效。MOFs 的这些特性使得其在未来药物递送系统中的应用前景广阔，为癌症、阿尔茨海默病及慢性病等疾病的治疗提供了新的可能性。未来的研究将继续探索 MOFs 在生物医学领域的应用潜力，开发出更加高效、安全和智能的药物递送系统。

1.2　为何用 MOFs 载药

近年来，随着纳米技术和材料科学的迅速发展，MOFs 在药物递送系统（DDS）

中的应用已逐渐成为科研领域的研究热点。基于传统的药物递送方法面临许多挑战，包括药物溶解性差、生物可用性低、靶向性差和副作用大等问题，MOFs 凭借其高孔隙率、可调节的孔道大小以及易于表面修饰的特性，为解决这些问题提供了新的思路。MOFs 的这些独特性质使它们能够在药物递送系统中扮演多种角色。首先，MOFs 可以高效载药，并通过调整 MOFs 的化学组成或物理结构来控制药物的释放速率，从而提高治疗效果。此外，MOFs 的表面可以进行功能化修饰，使得药物分子能够靶向病变组织或细胞，增加药物的局部浓度，从而进一步提高疗效和减少副作用。

1.2.1 MOFs 在药物递送领域的起源与现状

自从 Ferey 和其同事在 2006 年首次报道了金属有机框架材料 MIL-100 和 MIL-101 可用于装载布洛芬以来，科研工作者们已经设计和构建了各种基于 MOFs 的纳米平台，以满足药物递送领域的各种需求。基于 MOFs 的纳米平台在高效药物递送领域的进展，包括治疗癌症[21]、负载抗菌药物[22]、负载核酸和蛋白质[23]等。研究人员开发了许多 MOFs 作为药物载体，如 MIL-100[24]、ZIF-8[25]、MIL-53[26]、MOF-74[27]、Gd-MOF[28]、UiO-66（图 1.2）[29]。

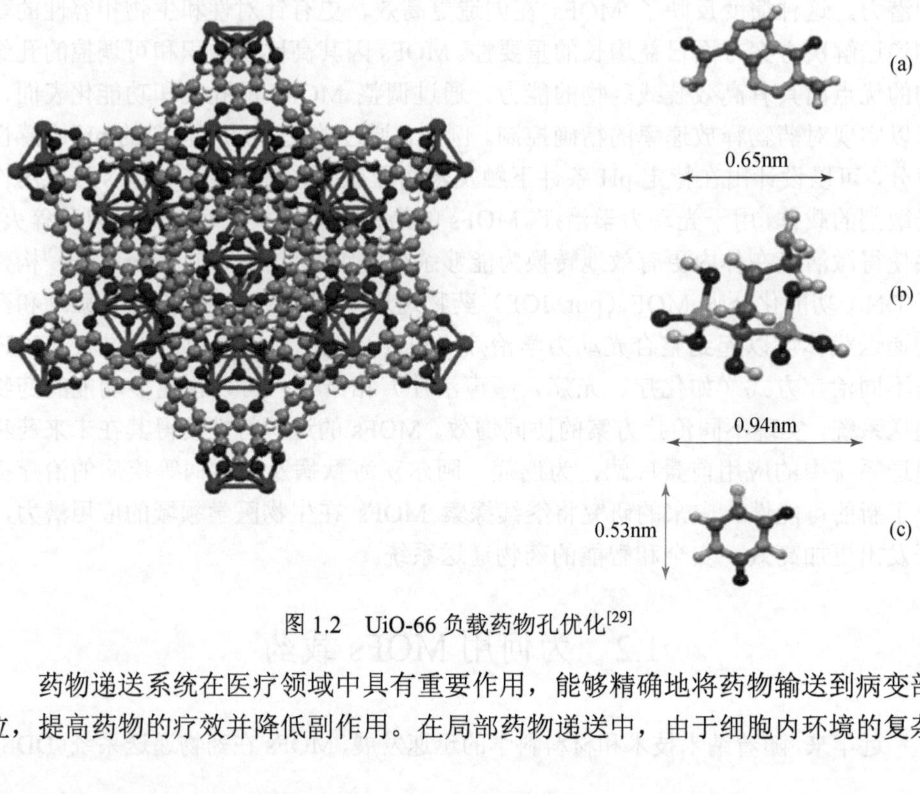

图 1.2　UiO-66 负载药物孔优化[29]

药物递送系统在医疗领域中具有重要作用，能够精确地将药物输送到病变部位，提高药物的疗效并降低副作用。在局部药物递送中，由于细胞内环境的复杂性，开发合适且可靠的平台进行可控的药物释放需求尤为强烈，实现可控药物释放将对解释细胞摄取的机制和指导新药的设计具有重要意义。例如 MIL-100 和 MIL-101 对布洛芬有较好的载药和释放效果，实验表明其固载率和缓释时间分别为 350 mg/g（3 天）以及 1400 mg/g（6 天）[24]。MOFs 作为一类新型材料，具有广泛的应用前景和巨大的发展潜力。未来的研究将围绕品种多样化、性能优化、合成方法创新和应用领域扩展展开，不断推动 MOFs 的发展。

近年来通过对微观形貌和化学组成的进一步改进，新型功能化 MOFs 问世并受到广泛关注。这些 MOFs 不仅继承了传统 MOFs 的优势，还表现出许多新型和增强的性能，如低密度、高负载能力和强催化活性等。从布洛芬的装载和控释开始，MOFs 在药物传递中的应用主要集中在小分子药物的释放上，如抗肿瘤药物（阿霉素和姜黄素）[30]。而随着研究的不断深入，已有研究使用 MOFs 封装和保护药物大分子（仿生物矿化）[31]。例如有研究开发了一种沸石咪唑酯框架（ZIF）纳米晶体的原位仿生矿化策略，构建具有良好细胞相容性、高稳定性和高 pH 响应性的药物释放体系[32]。将溶菌酶包裹在 ZIF（ZIF-8）表面，与金属离子紧密结合，促进骨样羟基磷灰石（HAp）的成核和生长，形成 HAp@Lys/ZIF-8 复合材料。体外实验表明，拥有空心 Lys/ZIF-8 核和 HAp 壳的复合材料具有较高的载药效率（56.5%）、良好的 pH 响应药物传递、细胞相容性和生理条件下的稳定性等优点。

小分子的装载递送一直是药物研究领域热点[33]，MOFs 具有显著的装载能力，允许各种具有不同物理化学性质的药物，如药物小分子[34]、寡肽，甚至蛋白质[35]的装载。MOFs 的孔隙率和成分可以通过仔细选择有机成分和金属离子来进行调整，以实现精确的物理和化学特征的构建。此外，适当的功能修饰可以使药物以可控和稳定的方式分布[36]。虽然报道的 MOFs 载药/固载系统多种多样[37-39]，但是报道最多的还是小分子药物或活性小分子成分（图 1.3）[40-45]。它们的配位键较弱，在生物条件下达到疗效后可迅速分解，这使得金属有机框架载药材料具有良好的生物降解性和生物相容性，甚至可负载核酸[46]。

1.2.2 MOFs 载药优势

金属有机框架（MOFs）的灵活组成、可调节的孔径大小和易于功能化的特性使其成为开发各种功能系统的关键组成部分，其在生物医学中的广泛研究已经在控制它的关键属性（如毒性、尺寸和形状以及生物稳定性）方面取得了实质性进展。

MOFs 在药物递送领域具有显著的潜力，主要得益于它们的独特载药特性。首先，MOFs 具备高比表面积和可调控的孔隙结构，这使得它们能够装载大量药物分子。Pamela 等通过使用不同比例的均苯三甲酸（BTC）与双氯芬酸（DCF）作为配体，把 DCF 直接纳入 MIL-100（Fe）框架，他们的研究发现孔隙度越高，药物释放量也越强[47]。

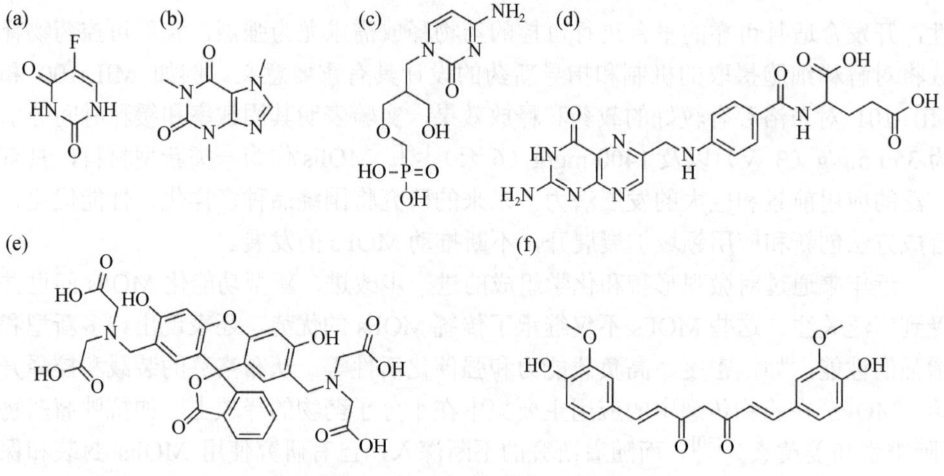

图 1.3　MOFs 装载小分子药物

（a）5-氟尿嘧啶[40]；（b）咖啡因[41]；（c）西多福韦[42]；（d）叶酸[43]；（e）钙黄绿素[44]；（f）姜黄素[45]

MOFs 粒径小、比表面积大、孔隙率高的特点使其能够装载和封装更多的药物分子。药物可以物理或化学吸附到 MOFs 的孔隙中，在那里可以长时间以受控的速率释放。MOFs 的孔径可以在合成过程中精确调整，从而可以封装特定尺寸的分子并控制它们的释放速率。这种孔隙结构的定制对于实现所需的释放动力学至关重要。MOFs 材料在药物递送领域的应用中，其独特的可调节孔隙结构和表面功能化能力，为实现药物的高效装载和释放提供了可能。此外，利用该特性，MOFs 的应用不仅限于传统药物的递送，还扩展到了生物大分子（如蛋白质和核酸等）的递送，进一步拓宽了其在药物递送领域的应用范围。

MOFs 的结构和特性可以被定制，这意味着它可以适应各种人体环境，并装载更多类型的药物。在结构方面，MOFs 的柔性或刚性孔隙形状（如孔道和笼）是多样的。另外，孔隙的大小，如微孔、介孔等，也可以得到控制。这样，不同形状和大小的药物分子就可以被放置在孔中。在性能上，MOFs 的吸附能力与框架的极性有关，因此在合成过程中或合成后，通常可以通过改变框架中的有机官能团来调整其吸附能力。另外，MOFs 的特殊结构可以与各种刺激相互作用。孔隙结构的存在使得 MOFs 在结构上普遍开放，从而导致底物和产物很容易从孔隙中转移，极大地促进了被封装的药物分子与环境刺激之间的相互作用。

对 MOFs 的化学功能化使得它们可以对药物释放进行更精确控制，例如，通过改变 MOFs 的化学组成或表面修饰，可以实现对药物释放环境（如 pH 值）的响应。ZIF-8 是一种由锌离子与 2-甲基咪唑配位形成，具有高负载和高 pH 敏感降解等独特特性的 MOF 材料。Xu 等[48]开发了一种组氨酸多肽杂交纳米金属有机框架材料，研究表明，该基于 ZIF-8 的载药系统在低 pH 条件下具有溶解和控制速率的独特性质。在 pH 值为 6.3 的 PBS 缓冲液中释放最快，4 h 时释放率达到 50%。

Sarker 等[49]研究了模型药物与活性炭 UiO-66 和 UiO-66-COOH 的吸附和释放，以了解功能化 MOFs 在药物传递和 MOFs 之间化学相互作用中的可能应用。结果表明，UiO-66-COOH 可作为药物存储/输送的有效材料，并可以通过调整 pH 值控制药物释放速率。

MOFs 的多功能性也是它们在药物递送系统中非常有价值的一个特点。通过表面修饰特定的靶向分子，可以使 MOFs 能够靶向特定类型的细胞或组织，从而提高药物的疗效同时减少对正常组织的副作用。如在癌症的治疗中可以用叶酸对 MOFs 进行表面修饰以使药物靶向肿瘤上的叶酸受体，使得药物能够通过细胞受体介导的方式被更有效地摄取。Zhong 等[50]采用模型抗原卵清蛋白（OVA）和 ZIF-8 一体化的方法合成了纳米颗粒（ZANPs），其高抗原装载能力达到 30.6%，并具有 pH 依赖的抗原释放性质，体内皮下注射后，ZANPs 可以靶向淋巴结（LNs）抑制 EG7-OVA 肿瘤小鼠的肿瘤生长。

MOFs 的另一个重要特性是它们的刺激响应性，这种特性使 MOFs 可以在特定的生物或化学信号存在时触发药物释放。与传统的药物递送系统相比，多模式疗法被认为是有前途的治疗策略，可以避免传统的药物递送系统在低负载量、不受控释放、非靶向或生物毒性方面具有的各种缺点。基于新型二维材料石墨炔（graphdiyne，GDY）与金属有机框架的杂交，Xue 等开发了多功能三维肿瘤靶向药物递送系统 Fe_3O_4@UiO-66-NH_2 / graphdiyne 结构，Fe_3O_4@UIO-66-NH_2(FU)[51]。合成的 FU-MOF 结构具有出色的磁性靶向能力，可以通过原位生长方法构建，在该方法中，它通过酰胺键与 GDY 表面结合，作为抗癌药物的载体，能够在肿瘤细胞周围有效释放药物。MOFs 作为新兴的多孔无机-有机杂化晶体材料，展示出比传统纳米药物载体更优越的性能。其框架结构有利于增加药物负载量，加快小分子扩散。

尽管纳米金属有机框架（NMOFs）在生物医药领域的发展仍处于初级阶段，但其在细胞靶向、药物递送、分子影像和肿瘤治疗方面已经展现出巨大潜力。预示着未来可以通过优化 MOFs 结构和性能，探索更广泛的医疗应用[52]。有研究报道用聚乙烯亚胺（PEI）包裹 ZIF-8，能够增强载体材料的负载能力和与 pDNA 的结合能力，在 MCF-7 细胞中观察到了良好的基因传递和表达[53]。MOFs 可通过响应各种疾病和病理的特殊刺激条件，实现精确和可控的药物释放，大大减少各种不良反应和副作用。Xie 等[54]把靶向试剂叶酸（FA）分子通过共价偶联的方式连接在装载 5-Fu 的纳米颗粒 MIL-101（Fe）-NH_2 的表面，细胞毒性试验表明，与 MIL-101（Fe）@Fu 和游离 5-Fu 相比，合成的纳米颗粒具有更好的生物相容性。此外，Dong 等[55]研究探索了一种 pH 响应聚合物修饰的多变量纳米 MOFs 药物传递系统，通过模板配体置换进行化疗或化学动力学联合治疗。采用一锅法合成

DOX@Cu/ZIF-8 并作为模板，在模板中用 3-AT 部分替换咪唑配体，获得 MAF@DOX。从生产到载药，从载药到释放和最终清除，MOFs 已被证明适合给药，目前已被用于有效封装小分子化疗药物和各种生物大分子。随着研究和技术进步，MOFs 有望成为集诊断与治疗于一体的多功能材料。

MOFs 在药物递送领域的重要性在于其独特的物理化学性质和灵活的功能化能力，能够满足特定药物递送系统的多样化需求。随着对 MOFs 结构和功能的深入理解，以及合成技术的不断进步，它将在未来的药物递送系统中发挥更加重要的作用。

参考文献

[1] Zhou Hong-Cai, Susumu K. Metal-organic frameworks (MOFs). Chemical Society Reviews, 2014, 43: 5415-5418.

[2] Xing Wendong, Yan Yulong, Wang Chong, et al. MOFs self-assembled molecularly imprinted membranes with photoinduced regeneration ability for long-lasting selective separation. Chemical Engineering Journal, 2022, 437: 135128.

[3] Liu Xiaomei, Tang Bing, Long Jilan, et al. The development of MOFs-based nanomaterials in heterogeneous organocatalysis. Science Bulletin, 2018, 63: 502-524.

[4] Sun Qianyu, Yuan Tianzhong, Yang Gang, et al. Chitosan-graft-poly(lactic acid)/Cd-MOFs degradable composite microspheres for sustained release of curcumin. International Journal of Biological Macromolecules, 2023, 253: 127519.

[5] Aguilera-Sigalat J, de Pipaón C S, Hernández-Alonso D, et al. A metal-organic framework based on a tetra-arylextended calix[4]pyrrole ligand: Structure control through the covalent connectivity of the linker. Crystal Growth & Design, 2017, 17: 1328-1338.

[6] Xu Jiong, Liu Jin, Li Zhen, et al. Optimized synthesis of Zr(Ⅳ) metal organic frameworks (MOFs-808) for efficient hydrogen storage. New Journal of Chemistry, 2019, 43: 4092-4099.

[7] Yang Min, Zhou Ya-Nan, Cao Yu-Ning, et al. Advances and challenges of Fe-MOFs based materials as electrocatalysts for water splitting. Applied Materials Today, 2020, 20: 100692.

[8] Yang Fan, Yang Guo-Ping, Wu Yunlong, et al. Ln(III)-MOFs (Ln=Tb, Eu, Dy, and Sm) based on triazole carboxylic ligand with carboxylate and nitrogen donors with applications as chemical sensors and magnetic materials. Journal of Coordination Chemistry, 2018, 71: 2702-2713.

[9] Guo Bing-Bing, Yin Jia-Cheng, Li Na, et al. Recent progress in luminous particle-encapsulated host-guest metal-organic frameworks for optical applications. Advanced Optical Materials, 2021, 9: 2100283.

[10] Huxford R C, Rocca J D, Lin Wenbin. Metal-organic frameworks as potential drug carriers. Current Opinion in Chemical Biology, 2010, 14: 262-268.

[11] Chen Yulu, Su Jiaqi, Dong Wenxia, et al. Cyclodextrin-based metal-organic framework nanoparticles as superior carriers for curcumin: Study of encapsulation mechanism, solubility, release kinetics, and antioxidative stability. Food Chemistry, 2022, 383: 132605.

[12] Forna N, Damir D, Duceac L D, et al. Nano-architectonics of antibiotic-loaded polymer particles as vehicles for active molecules. Applied Sciences, 2022, 12: 1998.

[13] Dougherty T J. Photodynamic therapy (PDT) of malignant tumors. Critical Reviews in Oncology/Hematology, 1984, 2: 83-116.

[14] Yang Weitao, Deng Cuijun, Shi Xiudong, et al. Structural and molecular fusion mri nanoprobe for differential diagnosis of malignant tumors and follow-up chemodynamic therapy. ACS Nano, 2023, 17: 4009-4022.

[15] Varshney A N, Vanidassane I, Ramavth D, et al. Chemotherapy in advanced thymic malignancies. Annals of Oncology, 2019, 30: 471P.

[16] Sousa J M D, Balbontín R, Durão P, et al. Multidrug-resistant bacteria compensate for the epistasis between resistances. PLOS Biology, 2017, 15: e2001741.

[17] Biserčić M S, Biserčić B, Zasońska B A, et al. Novel microporous composites of MOF-5 and polyaniline with high specific surface area. Synthetic Metals, 2020, 262: 116348.

[18] Nian Fuyu, Huang Yafan, Song Meiru, et al. A Novel fabricated material with divergent chemical handles based on Uio-66 and used for targeted photodynamic therapy. Journal of Materials Chemistry B, 2017, 5: 6227-6232.

[19] Pantwalawalkar J, Mhettar P, Nangare S, et al. Stimuli-responsive design of metal-organic frameworks for cancer theranostics: Current challenges and future perspective. ACS Biomaterials Science & Engineering, 2023, 9: 4497-4526.

[20] Zhao Qiu-ge, Zhou Yun-jie, Cao Dong-xiao, et al. DNA-functionalized porphyrinic metal-organic framework-based drug delivery system for targeted bimodal cancer therapy. Journal of Medicinal Chemistry, 2023, 66: 15370-15379.

[21] Wu Ming-Xue, Yang Ying-Wei, Metal-organic framework (MOF)-Bbased drug/cargo delivery and cancer therapy. Advanced Materials, 2017, 29: 1606134.

[22] Livesey T C, Mahmoud L A M, Katsikogianni M G, et al. Metal-organic frameworks and their biodegradable composites for controlled delivery of antimicrobial drugs. Pharmaceutics, 2023, 15: 274.

[23] Wang Shunzhi, Chen Yijing, Wang Shuya, et al. DNA-functionalized metal-organic framework nanoparticles for intracellular delivery of proteins. Journal of the American Chemical Society, 2019, 141: 2215-2219.

[24] Mileo P G M, Gomes D N, Gonçalves D V, et al. Mesoporous metal-organic framework mil-100(Fe) as drug carrier. Adsorption, 2021, 27: 1123-1135.

[25] Zhang Huiyuan, Li Qian, Liu Ruiling, et al. A versatile prodrug strategy to in situ encapsulate drugs in MOF nanocarriers: A case of cytarabine-Ir820 prodrug encapsulated ZIF-8 toward chemo-photothermal therapy. Advanced Functional Materials, 2018, 28: 1802830.

[26] Li Anxia, Yang Xiaoxin, Chen Juan. A novel route to size-controlled MIL-53(Fe) metal-organic frameworks for combined chemodynamic therapy and chemotherapy for cancer. RSC Advances, 2020, 11: 10540-10547.

[27] Hu Jiaqi, Chen Yi, Zhang Hui, et al. Tea-assistant synthesis of MOF-74 nanorods for drug delivery and *in-vitro* magnetic resonance imaging. Microporous and Mesoporous Materials, 2020, 315: 110900.

[28] Li Li, Zhu Yueming, Qi Zhaorui, et al. Gcmc simulations to study the potential of MOFs as drug delivery vehicles. Applied Organometallic Chemistry, 2023, 37: e7199.

[29] Boroushaki T, Koli M G, Malekshah R E, et al. Elucidating anticancer drugs release from UiO-66 as a carrier through the computational approaches. RSC Advances, 2023, 13: 31897-31907.

[30] Ma Wenzhuan, Guo Qiang, Li Ying, et al. Co-assembly of doxorubicin and curcumin targeted micelles for synergistic delivery and improving anti-tumor efficacy. European Journal of Pharmaceutics and Biopharmaceutics, 2017, 112: 209-223.

[31] Wang Yueqing, Ma Jizhen, Cao Xueying, et al. Bionic mineralization toward scalable MOF films for ampere-level biomass upgrading. Journal of the American Chemical Society, 2023, 145: 20624-20633.

[32] Huang Ting, Yang Ling, Wang Shuqiang, et al. Enhanced performance of ZIF-8 nanocrystals hybrid monolithic composites *via* an *in-situ* growth strategy for efficient capillary microextraction of perfluoroalkyl phosphonic acids. Talanta, 2023, 259: 124452.

[33] Ji Yu, Zhu Ruiyao, Shen Yue, et al. Comparison of loading and unloading of different small drugs on graphene and its oxide. Journal of Molecular Liquids, 2021, 341: 117454.

[34] Motakef-Kazemi N, Shojaosadati S A, Morsal A. *In situ* synthesis of a drug-loaded MOF at room temperature. Microporous and Mesoporous Materials, 2014, 186: 73-79.

[35] Díaz J C, Giménez-Marqués M. Alternative protein encapsulation with MOFs: Overcoming the elusive mineralization of HKUST-1 in water. Chemical Communications, 2024, 60: 51-54.

[36] Tan Yan-Xi, He Yan-Ping, Zhang Jian. Tuning MOF stability and porosity *via* adding rigid pillars. Inorganic Chemistry, 2012, 51: 9649-9654.

[37] Du Jinsong, Chen Guanping, Yuan Xinyi, et al. Multi-stimuli responsive Cu-MOFs@Keratin drug delivery system for chemodynamic therapy. Frontiers in Bioengineering and Biotechnology,

2023, 11: 1125348.

[38] Rather R A, Siddiqui Z N. Silver phosphate supported on metal-organic framework (Ag_3PO_4@MOF-5) as a novel heterogeneous catalyst for green synthesis of indenoquinolinediones. Applied Organometallic Chemistry, 2019, 33: e5176.

[39] Hasanzadeh M, Hashemzadeh N, Shadjou N, et al. Sensing of doxorubicin hydrochloride using graphene quantum dot modified glassy carbon electrode. Journal of Molecular Liquids, 2016, 221: 354-357.

[40] Hu Zengchi, Qiao Chengfang, Xia Zhengqiang, et al. A luminescent Mg-metal-organic framework for sustained release of 5-fluorouracil: Appropriate host-guest interaction and satisfied acid-base resistance. ACS Applied Materials & Interfaces, 2020, 12: 14914-14923.

[41] Devautour-Vinot S, Martineau C, Diaby S, et al. Caffeine confinement into a series of functionalized porous zirconium MOFs: A joint experimental/modeling exploration. The Journal of Physical Chemistry C, 2013, 117: 11694-11704.

[42] Murad F, Minasny B, Bramley H, et al. Development of a crop water use monitoring system using electromagnetic induction survey. Soil & Tillage Research, 2022, 223: 105451.

[43] Lin Caixue, Chi Bin, Xu Chen, et al. Multifunctional drug carrier on the basis of 3d-4f Fe/La-MOFs for drug delivery and dual-mode imaging. Journal of Materials Chemistry B, 2019, 7: 6612-6622.

[44] Ibrahim M, Sabouni R, Husseini G A, et al. Facile ultrasound-triggered release of calcein and doxorubicin from iron-based metal-organic frameworks. Journal of Biomedical Nanotechnology, 2020, 16: 1359-1369.

[45] Lawson S, Siemers A, Kostlenick J, et al. Mixing Mg-MOF-74 with Zn-MOF-74: A facile pathway of controlling the pharmacokinetic release rate of curcumin. ACS Applied Bio Materials, 2021, 4: 6874-6880.

[46] Sun Yuqing, Yu Haixin, Han Shaoqing, et al. Method for the extraction of circulating nucleic acids based on MOF reveals cell-free RNA signatures in liver cancer. National Science Review, 2024, 11: nwac022.

[47] So P B, Chen H-T, Lin C-H, De Novo synthesis and particle size control of iron metal organic framework for diclofenac drug delivery. Microporous and Mesoporous Materials, 2020, 309: 110495.

[48] Xu Yanan, Li Zhenhua, Xiu Dan, et al. Histidine polypeptide-hybridized nanoscale metal-organic framework to sense drug loading/release. Materials & Design, 2021, 205: 109741.

[49] Sarker M, Jhung S H. Zr-MOF with free carboxylic acid for storage and controlled release of caffeine. Journal of Molecular Liquids, 2019, 296: 112060.

[50] Zhong Xiaofang, Zhang Yunting, Tan Lu, et al. An aluminum adjuvant-integrated nano-MOF as antigen delivery system to induce strong humoral and cellular immune responses. Journal of Controlled Release, 2019, 300: 81-92.

[51] Xue Zhongbo, Zhu Mengyao, Dong Yuze, et al. An integrated targeting drug delivery system based on the hybridization of graphdiyne and MOFs for visualized cancer therapy. Nanoscale, 2019, 11: 11709-11718.

[52] Mao Xuanxiang, He Fangni, Qiu Dehui, et al. Efficient biocatalytic system for biosensing by combining metal-organic framework (MOF)-based nanozymes and G-quadruplex (G4)-dnazymes. Analytical Chemistry, 2022, 94: 7295-7302.

[53] Li Yantao, Zhang Kai, Liu Porun, et al. Encapsulation of plasmid DNA by nanoscale metal-organic frameworks for efficient gene transportation and expression. Advanced Materials, 2019, 31: 1901570.

[54] Xie Wen, Zhou Feiya, LiXiang, et al. A surface architectured metal-organic framework for targeting delivery: Suppresses cancer growth and metastasis. Arabian Journal of Chemistry, 2022, 15: 103672.

[55] Dong Junliang, Yu Yueyuan , PeiYuxin, et al. pH-responsive aminotriazole doped metal organic nrameworks nanoplatform enables self－boosting reactive oxygen species generation through regulating the activity of catalase for targeted chemo/chemodynamic combination therapy. Journal of Colloid and Interface Science, 2022, 607: 1651-1660.

第2章 金属有机框架材料的制备

MOFs 的性能在很大程度上取决于其精确的化学组成和结构特征，而这些特征则由其制备方法决定。因此，开发有效、可控的 MOFs 合成策略成为该研究领域的核心任务之一。

随着对 MOFs 结构与性能关系认识的深入，传统的合成方法如溶剂热法已经被逐渐优化并扩展，新兴的合成策略如机械化学法、微波辅助合成法等也相继被开发，这些方法在提高合成效率、降低成本和环境影响方面展现出独特的优势。随着研究的不断深入，MOFs 制备技术的发展也面临着一系列挑战，主要包括如何实现 MOFs 结构的精确控制、提高材料的稳定性和功能化程度以及满足实际应用中对材料性能的要求等问题。本章旨在提供一个全面的 MOFs 制备技术概述，包括各种合成方法的比较分析、挑战讨论以及未来发展趋势的预测，为 MOFs 的进一步研究和应用提供理论基础和技术支持。通过不断改进 MOFs 的制备方法，期望能够促进金属有机框架材料在载药领域的发展，推动 MOFs 材料在药物递送应用领域的突破。

2.1 MOFs 的物理化学性质

MOFs 的组成可以分为四级：第一级是用于构建 MOFs 的化学成分，即单价或多价金属离子（节点）和配位配体（连接体）。其中锆（Ⅳ）、铁（Ⅲ）和锌（Ⅱ）是 MOFs 中最常用的离子。第二级中用于 MOFs 合成的配体通常具有多个羧基或胺类官能团，它们从烷基链或苯或咪唑等环基结构延伸出来。配体与离子的配位形成了具有规则重复几何形状的晶体状晶格。第三级组成被称为二级构筑单元(SBU)，它是多个配体与金属离子形成相对刚性的几何结构 SBU 作为 MOFs 结构生长的模板或单位细胞，通过桥接配体连接金属节点，定义了 MOFs 的内部框架和孔隙结构。在 MOFs 的前三个结构水平中，协调金属和配体的选择和排列决定了 MOFs 的外部形态。具体而言，桥接配体在金属节点之间形成连接，使得 MOFs 的内部框架得以稳定，并生成丰富的孔隙结构。在第四级结构水平上，MOFs 的外部形态则受到合成方法和合成过程中是否封装分子（如治疗药物）的影响。不同的合成方法可以调控 MOFs 的晶体尺寸、形态以及孔隙特性，从而影响其在实际应用中的表现。

此外，MOFs 含有配位不饱和金属位点，它可以像 Lewis 酸一样发挥作用，帮助将分子加载到表面并使 MOFs 功能化。对 MOFs 的化学和结构特征进行精细、多水平的控制，可使其很好地应用到药物传递中。MOFs 的结构决定了其是否具有优异的物理化学性能，包括易于合成和功能化、可定制的孔隙大小、可变结构、高表面积和加载能力、良好的生物相容性和生物降解性等。

2.1.1 比表面积

比表面积和孔隙率是表征 MOFs 的最重要的量，通过 BET 分析可以从 N_2 吸附等温线中测定 MOFs 材料比表面积。一般来说，MOFs 吸附的形成不是通过层的形成，而是遵循孔隙填充机制，并表现出极大的 BET 比表面积。Walton 等研究了 BET 法测定 MOFs 比表面积的适用性。其研究表明，从模拟等温线测量到的 BET 比表面积与直接从几何模式下的晶体结构确定的比表面积吻合良好，并与文献中的实验报告结果一致，这项结果进一步证实了使用 BET 理论来获得 MOFs 的比表面积[1]方法的可行性。

MOFs 的高比表面积显著增强了其载药能力，允许在其结构中吸附更多的药物。此功能对于开发高效的药物递送系统至关重要，因为它能够输送足够的药物剂量，同时最大限度地减少给药的大小和频率，从而可能减少副作用并提高患者的依从性。此外，高比表面积有助于增加与生物分子的相互作用，可以针对靶向药物释放和提高生物利用度进行定制。MOFs 材料的比表面积常在数千平方米每克以上，和传统的分子筛和活性炭等多孔材料相比性能优越。科研人员把锌离子和多齿羧酸配位，制备了一系列超高比表面积的 MOFs 材料，其中一些金属有机框架材料的比表面积高达 6240 m^2/g。Zhou 等研制的 PCN 系列材料的比表面积最高可达 5109 m^2/g；Arne 等成功合成了 NU-110 材料，其比表面积达到了惊人的 7140 m^2/g[2, 3]。

MOFs 材料中 MOF-5（$Zn_4O(BDC)_3$）通常具有约 3000 m^2/g 的比表面积，是研究中的一个常见比表面积基准[4]。HKUST-1（$Cu_3(BTC)_2$）[5]比表面积可达约 2200 m^2/g。UiO-66 具有约 1000～1600 m^2/g 的比表面积，具体取决于后处理方法和合成条件[6]。MOF-177（$Zn_4O(BTB)_2$）[7]是已知比表面积最大的 MOFs 之一，可达到 4500 m^2/g。此外，MIL-101(Cr)[8]同样具有较高的比表面积，可以达到约 3100 m^2/g。ZIF-8（$Zn(mIm)_2$）的比表面积通常在 1000～1500 m^2/g，根据合成条件的不同而有所变化[9]。COF-105 作为共价有机框架材料，它的比表面积超过 4000 m^2/g，说明了多孔材料在比表面积上的极限[10]。PCN-777[11]是一种具有开放金属位点的 MOF，其比表面积可达到 4500 m^2/g。

高比表面积造就了 MOFs 优异的载药量，李婉萌等[12]合成了一种采用金属有机框架（Cu-MOF）作为载体材料负载 BRH（一种异喹啉类生物碱盐酸小檗碱）的材料。结果表明 Cu-MOF@BRH 的载药量为 48%±2.41%，缓释作用明显，48 h 累计释放率为 53%。王秋月等采用"一锅法"包载 5-氟尿嘧啶（5-fluorouracil，5-Fu）制备 5-Fu@ZIF-8 载药纳米粒[13]，结果显示两种纳米粒具有良好的载药特性，5-Fu@ZIF-8 和 5-Fu@ZIF-8@PDA 的包封率分别为 79.55%±5.50%和 98.41%±1.61%（图 2.1）。

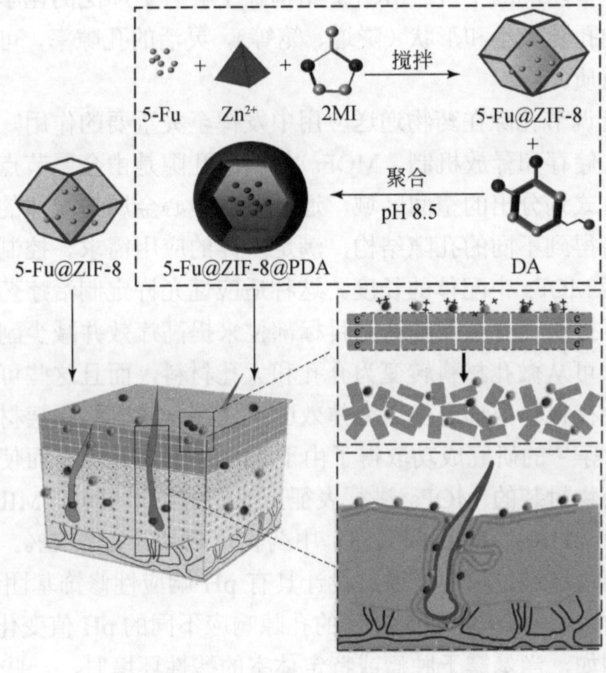

图 2.1　"一锅法"包载 5-氟尿嘧啶[13]

Guo 等[14]设计了一种 MOF，首先将 MTX 与单宁酸（TA）按 2∶1 的比例偶联后与铁离子（Fe^{3+}）配位，然后对透明质酸（HA）进行表面修饰。研究结果表明，得到的 MOFs 达到超高的载药量（45%）和持续的释药特性，可选择性识别病变细胞达到抗炎作用。Lin 等[15]合成的一种新型的核壳 MOFs/CDs@OCMC 纳米颗粒，可用于癌症的体外诊断和治疗。以阿霉素（Dox）作为模型药物，载药量为 50 mg/g。Nasrabadi 等用溶剂热法制备出 UiO-66，并将环丙沙星（CIP）装入 UiO-66 中，检测其载药量和抗菌活性[16]。研究结果显示，UiO-66 具有非常高的 CIP 装载率，高达 84%，对金黄色葡萄球菌和大肠杆菌具有显著的抑菌活性。

MOFs 的高比表面积允许其载入大量的药物分子，为高效药物递送系统的构建提供了可能。这一特性使得 MOFs 可以在医药领域，尤其是在靶向治疗和控制释放系统中发挥重要作用。

2.1.2 可调节孔隙

MOFs 的可调节孔隙允许封装各种药物分子，提供受控的药物释放机制。这一特性有利于创建能够响应特定生理条件（如 pH 值或温度）的药物递送系统，以在体内的目标部位释放药物。具有较大孔隙的 MOFs 可以封装更大的分子或更高的药物量，而较小孔隙的 MOFs 适用于较小的分子，同时又为亲水性和疏水性药物提供了一个多功能的平台。MOFs 结构除了具有更广泛的化学多功能性外，还表现出多种的孔隙大小和形状（隧道、笼等），灵活的孔隙率，可以使孔隙大小可逆地适应吸附质。

MOFs 的可调节孔隙在药物递送应用中发挥至关重要的作用，关系到精确控制药物的封装、储存和释放机制。MOFs 材料的孔隙是由金属节点和有机配体通过特定的连接方式划分出的空间区域，通过改变中心金属与配体的种类，采用不同的连接方式可得到不同的孔隙结构，满足多样的应用需求。控制孔隙大小最直接的方法是改变 MOFs 中配体的长度。这种适应性允许定制治疗药物的释放，通过确保药物以可控的速度直接输送到目标部位来提高疗效并减少副作用。因此通过调控，MOFs 可从微孔材料转变为介孔和大孔材料，而且这些可调节孔隙在符合一定条件时会在材料内部产生主客体效应，这在金属有机框架材料载药方面至关重要。Yadav 等[17]的研究成功取得了由于 MOFs 的高孔隙率而使其具有高载药量的进展。对药物封装的 MOFs 进行表征，他们的结果显示，MIL-100（Fe）中药物负载量约为 23.1%，MIL-Fe（53）中药物负载量约为 27.6%。

此外，在药物递送过程中，通过设计具有 pH 响应性修饰基团的 MOFs 或将 pH 敏感官能团掺入其结构能使 MOFs 的孔隙响应不同的 pH 值变化，从而实现靶向释放机制。例如，当暴露于肿瘤或特定体室的酸性环境时，一些 MOFs 会发生结构变化，改变其孔隙开口，以此来完成封装药物的释放。pH 响应行为通过确保治疗剂直接在疾病部位释放，从而最大限度地提高药物疗效。Xiong 等[18]开发了一种 pH 响应金属有机框架（MOFs）系统，由透明质酸（HA）修饰，装载抗炎症原儿茶酸（PCA）用于治疗骨关节炎。在酸性条件下中心离子 Fe^{3+}离子会发生降解，是一种 pH 诱导的 MIL-100（Fe）给药系统。MOF@HA@PCA 可对 OA 微环境中的酸性条件做出良好的反应，并逐渐释放 PCA，从而显著降低 IL-1β 诱导的软骨细胞和 OA 关节的滑膜炎症。

Pandey 等[19]将乳铁蛋白（Lf）用作蛋白质基质，用来装载钛茂，然后将其与 5-Fu 一起封闭在 ZIF-8 框架中。其药物的 pH 响应释放和 Zn^{2+}离子在细胞和分子水平上的调节对癌细胞具有有效的杀伤力。此外，Asadollahi 等[20]开发了海藻酸钠–玉米醇溶蛋白（SA-ZN）纳米复合材料改性金属有机框架，并将其作为柳氮磺胺吡啶（SSZ）传递的有效纳米载体，MOF/SA-ZN 纳米载体对于结肠传递过程具有有效促进作用。因此，可以利用 SA-ZN 涂层的 Zn-MOF 作为一个理想的平台来制备具有 pH 响应特性的治疗类风湿性关节炎的递送系统。

2.1.3　多功能结构：富含不饱和位点

MOFs 的多功能结构灵活性在药物递送领域非常重要，可定制框架设计，能满足特定的药物递送需求，包括靶向递送、控释和增强药物载量等。当这种适应性能够将各种功能集成到一个系统中时，即可同时完成给药、生物成像和治疗监测等功能，推进个性化医疗方法深入。通过表面修饰或化学改性，MOFs 可实现高度特定的药物释放，也可在增强治疗效果的同时减少副作用。

在 MOFs 的合成过程中，中心金属不仅与有机配体进行配位，还与溶剂分子配位以满足金属离子的配位需求。然而，这种配位并不稳定，加热或真空处理等方式会除去配位的溶剂分子，暴露中心金属的不饱和配位点[21]。不饱和金属位点具有高度的活性，能够与药物分子形成较强的配位键，从而增强药物分子在 MOFs 中的吸附能力。这意味着 MOFs 可以高效地装载更多的药物分子，提高药物递送系统的载药量。通过对不饱和金属位点的化学修饰，可以调控药物分子与 MOFs 之间的相互作用强度，进而实现对药物释放速率的精细调控。例如，通过改变 MOFs 中金属离子的种类或引入不同的有机配体，可以调节药物释放的 pH 敏感性或温度响应性，从而实现靶向释放和控制释放。这同时也为 MOFs 提供了一种方便的途径来引入功能性基团，使 MOFs 不仅能够作为药物载体，还可以通过表面修饰实现特定的生物识别、成像或治疗功能。通过在不饱和金属位点上引入靶向配体或诊断剂，MOFs 可以被赋予靶向特定细胞或组织的能力，增强药物递送的特异性和效率。例如叶酸部分被用于专门靶向几种叶酸受体（FR）的癌症。Jiang 等研制了一种基于沸石咪唑酸盐框架（ZIF-8）涂层 Au@Ag 核壳纳米棒的表面增强拉曼散射（SERS）成像探针和药物载体[22]。通过将叶酸固定在 Au@Ag NRs4-ATP@ZIF-8 上，实现将 SERS 探针应用于 HeLa、MCF-7、LNCaP、QGY-7703、HCT116 和 MDA-MB-231 细胞的靶向 SERS 成像。Ahmed 将 Dox 负载的 MOFs 与聚乙二醇叶酸（PEG-FA）进一步功能化，实现肿瘤上叶酸受体的靶向治疗[23]。

不饱和金属位点还可以与药物分子之间产生特定的相互作用（如氢键、π-π 相互作用等），这不仅有助于稳定药物分子，还可以促进药物在生理环境中的稳定释放，减少药物的非特异性释放和副作用。Cao 等通过 PEG 修饰 MOF 装载氨磷汀可以起到改善氨磷汀口服性差的作用[24]。通过在不饱和金属位点上引入具有生物相容性的配体或对其进行修饰，可以提高 MOFs 在生物体内的稳定性和生物相容性，使其更适合于生物医学应用。

2.1.4 改性及功能化

MOFs 的表面可以通过化学修饰进行功能化，以增强其与药物分子的相互作用，提高靶向性和生物相容性（图 2.2），还可以利用亲水配体的表面功能化提高 MOFs 的稳定性[25]。MOFs 材料可以通过多种策略进行改性和功能化，包括改变中心离子和有机配体的种类、修饰有机配体以及调整配位作用等。基于特殊需求，我们还可以采取直接加入多种金属离子或通过金属离子交换的方法，构建出多金属 MOFs。同样，含多种配体的 MOFs 可以通过引入不同的配体或采用配体交换

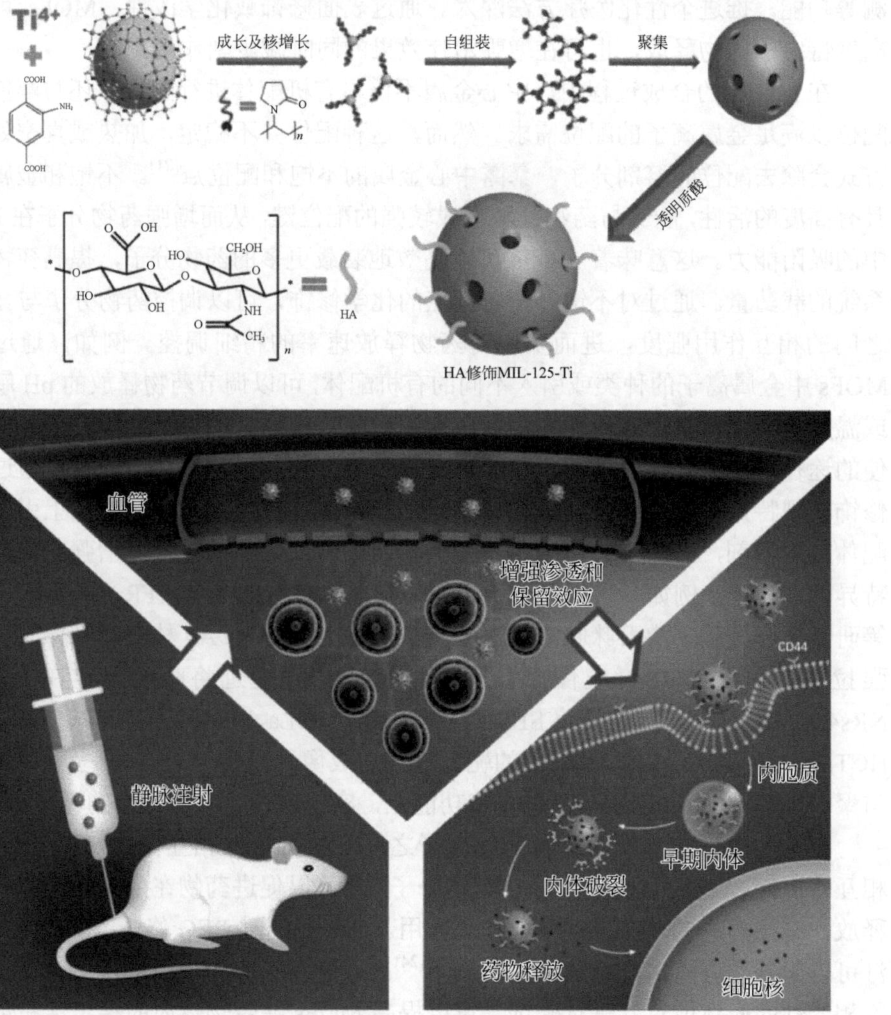

图 2.2　透明质酸（HA）修饰的 Ti 基 MOFs[26]

法制得[27]。而且，MOFs 的功能化可以直接使用带有功能性基团的配体来合成 MOFs[26]。

到目前为止，虽然关于容器分子的主客体化学已有大量研究，但在设计具有可控客体吸收和释放能力的纳米多孔材料方面仍存在许多不足。宿主和客体之间的相互作用通常难以有效改变，这意味着客体对宿主的亲和力是固定的，客体的进出空腔是基于一个简单的平衡，这种行为暂时无法得到有效控制。近年来，除了对多孔材料框架进行修饰外，在纳米多孔材料表面修饰一些高分子材料以提高其在生物体中的稳定性和生物相容性的方法也备受关注。

例如，Tan 等提出了一种将金属有机框架（MOFs）与热敏水凝胶集成的混合纳米复合材料，设计了一种可注射的植入物。将阿霉素（Dox）和塞来昔布（Cel）加入系统，用于局部口腔癌治疗（Dox/Cel/MOFs@Gel）。MOFs 和水凝胶的混合可以实现双重药物的稳定递送，抑制药物的突发性泄漏，并扩散到外围。Dox/Cel/MOFs@Gel 对 Dox 和 Cel 具有较高的包封能力和有效的 pH 响应特性，并提高了局部肿瘤治疗的化疗效率[28]。Alves 等[29]用 N_3-Bio-MOF-100 装载一种天然化合物 CCM，表面通过点击化学与 FA 基团功能化，可以实现特异性靶向癌细胞。Hu 等[30]构建了 γ-环糊精（CD）-MOFs 晶体，并将其改性为布地奈德（Bud）的 DPI 载体，使用胆固醇（Cho）和亮氨酸（Leu）-泊洛沙姆对 CD-MOFs 粉末进行改性，以改善流动性和颗粒空气动力学行为。与 CD-MOFs 相比，胆固醇修饰后肺组织的荧光信号得到明显改善。

通过简单的化学反应进行 MOFs 的表面和孔隙的功能化修饰，以引入特定的功能基团，这类功能化可以实现对药物分子的特定结合，增强药物的稳定性或改善其释放特性。例如，可以通过修饰 pH 敏感或温度敏感的官能团来制造能够在特定体内环境下释放药物的智能药物递送系统。通过在 MOFs 的表面引入特定的靶向配体（如抗体、蛋白质或小分子），可以增强 MOFs 对病变组织或特定细胞类型的亲和力，从而提高药物的靶向性和治疗效率。这种靶向性递送减少了药物对正常组织的副作用并提高了疗效。通过化学修饰，可以在同一 MOFs 结构中实现多种药物的分层或有序释放，这对于需要联合治疗的疾病（如多药耐药性肿瘤）具有特别价值。通过引入生物相容和可生物降解的有机链或其他生物相容性材料，可以改善 MOFs 在生物体内的行为，如减少潜在的毒性并增加其在体内的降解，这使得 MOFs 成为更安全的药物递送载体。"可改性"使其能够集成"治疗诊断一体化"策略，例如，可以在 MOFs 结构中同时装载药物和成像剂。这种集成有助于在治疗过程中实时监测药物的分布和释放情况。

MOFs 的易于改性不仅提高了其在药物递送领域的适应性和灵活性，还大大扩展了其在现代医疗中的潜在应用范围，为发展新型高效的药物递送系统提供了广阔的可能性。

2.2　MOFs 的分类

MOFs 独特的性质源于其多孔结构、高比表面积以及可调控的化学组成。通过改变 MOFs 的组成元素、有机配体以及组装方式，可以得到结构和功能各异的 MOFs 材料（表 2.1）。对 MOFs 进行有效的分类，不仅有助于理解其结构与性能之间的关系，也为材料的设计和应用提供了理论基础。

表 2.1　不同 MOFs 的载药应用[31-37]

MOFs	递送药物	相关应用
MOF-53（Fe）	BDC	抗菌
ZIF-8	头孢他啶	
ZIF-8	四环霉素	
ZIF-8	槲皮素	肿瘤
ZIF-8	阿霉素	白血病
GOx/ZIF-8	胰岛素	Type I DM
γ-CD-MOF	来氟米特	类风湿性关节炎
CD-MOF	阿齐沙坦	高血压
CD-MOF	姜黄素	肺部给药
Zr-MOF	5-氟尿嘧啶	癌症

MOFs 的分类通常基于其组成的金属节点、有机配体类型以及它们之间的连接方式。例如，根据中心金属的不同，MOFs 可以分为过渡金属 MOFs、稀土金属 MOFs 等；根据有机配体的不同，又可以分为含羧酸配体 MOFs、含 N 杂环配体 MOFs 等。此外，MOFs 还可以按照其空间结构和孔隙特性进行分类，如按照孔隙大小分为微孔 MOFs、中孔 MOFs 和大孔 MOFs；按照其三维网络结构的不同，可以分为立方体、八面体等多种几何形态的 MOFs。探讨不同分类方法及其背后的理论基础，通过具体实例了解不同类别 MOFs 的典型特性和应用场景。对 MOFs 进行系统分类，不仅可以更好地理解现有的材料体系，还能够指导未来新型 MOFs 结构的设计和合成，进一步拓宽其在药物递送中的应用。常用载药 MOFs 如图 2.3 所示。

2.2.1　IRMOFs

网状金属有机框架材料（isareticular metal-organic frameworks，IRMOFs）是由分离的次级结构单元$[Zn_4O]^{6+}$无机基团与一系列芳香羧酸配体键合而成的重复网络拓扑结构，是一种八面体形式桥连自组装形成的微孔晶体材料。这种材料具有较大的孔穴及孔容。与此同时，羧酸基有机配体更倾向于和金属离子形成簇，从而一定程度上阻止 MOFs 间的相互贯穿。目前常用于药物递送的 IRMOFs 有 IRMOF-1、IRMOF-8、IRMOF-10 和 IRMOF-16 等。

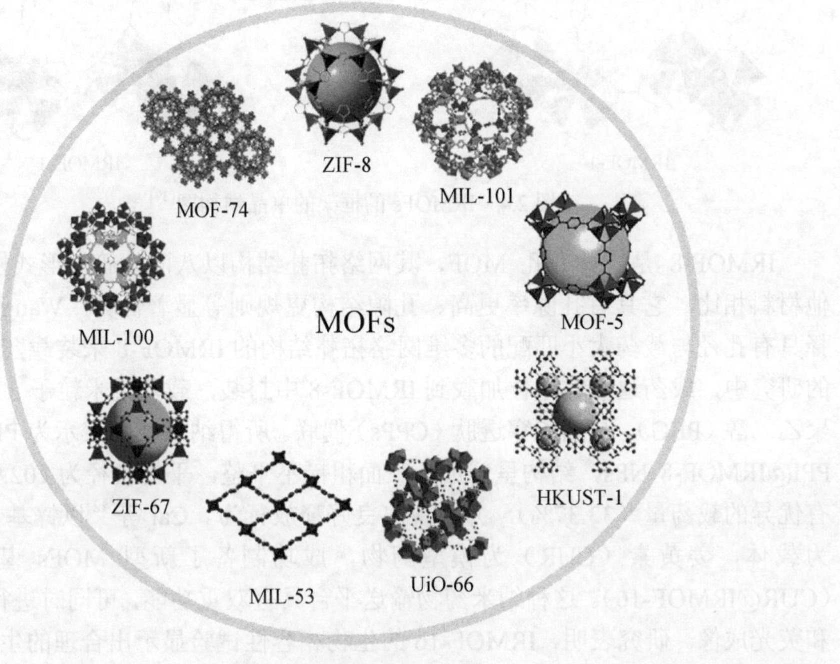

图 2.3　常用载药 MOFs[31]

在 IRMOFs 系列中，MOF-5（$ZnO_4(BDC)_3(DMF)_8C_6H_5Cl$，DMF 为二甲基甲酰胺）是最常见的一种，由 Yaghi 等于 1999 年首次合成出来[38]，使用对苯二甲酸（terephthalic acid）为有机配体与 Zn 离子通过八面体形式配位形成三维立体材料（图 2.4）。MOF-5 的比表面积可高达 2900 m^2/g，改善合成方法后甚至可达 3800 m^2/g。在后续的研究中，研究人员使用对苯二甲酸类似物将类 MOF-5 的孔径从 0.38 nm 提高至 2.88 nm，实现了中孔 MOFs 的合成。

IRMOF-3 也称为 NH_2-BDC-MOF-5，Xu 等合成了 Fe_3O_4@C@IRMOF-3-FA/PEG 多功能纳米载体，该载体具有靶向性以及微环境响应性。纳米载体可以在 360 nm 激发下发射蓝色荧光，实现光学监测细胞内化过程，并且表现出 74.9%的高负载药物能力[39]。Laha 等[40]开发了一种通用的合成策略，将天然分子姜黄素封装在单分散的等网状纳米级金属有机框架粒子（NMOF-3）中，以便在三阴性乳腺癌细胞（TNBC）中进行靶向药物递送。在体内研究中，与接受非靶向递送的小鼠相比，接受姜黄素靶向递送的小鼠具有更高的生存能力并减小了肿瘤体积。叶酸偶联姜黄素负载 IRMOF-3 可能对治疗 TNBC 有效。

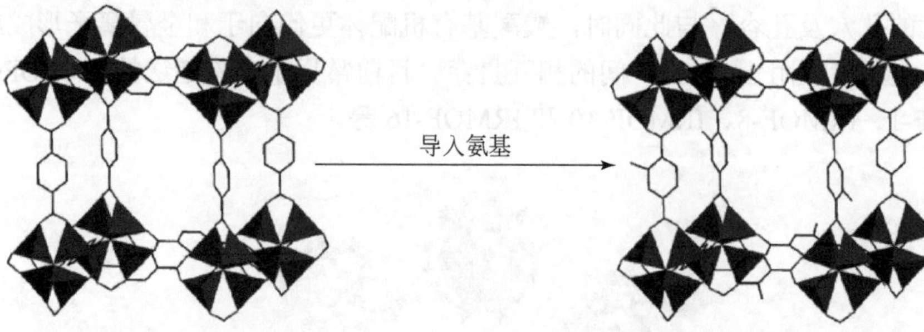

图 2.4 IRMOFs 的框架的单晶结构图[39]

IRMOF-8 是一种介孔 MOF，其网络拓扑结构以八面体簇的形式桥接。与其他材料相比，它具有孔隙率更高、孔隙结构更规则等显著优势。Wang 等[41]曾选择具有孔径与载药大小匹配的多维网络拓扑结构的 IRMOF-8 来装载药物。在他们的研究中，载药通过非原位加载到 IRMOF-8 中封装，载药纳米粒子（NPs）涂有聚乙二醇（PEG），与细胞穿透肽（CPPs）偶联。所得纳米载体表示为 PEG-CPP44/PPI@IRMOF-8 NPs，结构呈球形，表面粗糙不平整，平均粒径为 202.97 nm，具有优异的载药量（33.37%），并表现出良好释放行为。Cai 等[42]以锌基 IRMOF-16 为载体，姜黄素（CUR）为模型药物，成功制备了新型 MOFs 基纳米颗粒（CUR@IRMOF-16）。这种纳米药物输送平台具有双重功能，可同时进行药物输送和荧光成像。研究表明，IRMOF-16 的生物相容性试验显示出合理的生物安全性，体外未观察到明显毒性。

Kotzabasaki 等[43]研究调查了存储在金属有机框架 IRMOF-74-Ⅲ和功能化 OH-IRMOF-74-Ⅲ中的抗癌药物吉西他滨（Gem）的微观行为。准确的量子力学计算表明，在两种结构中 Gem-MOF 相互作用均可用于药物的吸附和缓慢释放。基于经典的 Monte Carlo 模拟模型，Gem 的预测最大负载量是脂质包覆的介孔二氧化硅纳米颗粒的 3 倍。此外，目前还有研究合成了共掺杂率高达 60%的共掺杂 Zn-MOF-5 纳米颗粒，用于化疗-化学动力学协同治疗肿瘤。Co 离子可以通过 Fenton 样反应介导化学动力学治疗，并通过消耗还原性谷胱甘肽来调节肿瘤微环境。CoZn-MOF-5 具有较高的载药能力，对阿霉素的载药率为 72.8%。且 CoZn-MOF-5 @PEG@DOX 纳米药物对 4T1 癌细胞具有较强的杀伤作用，促进了化疗-化学动力学协同作用[44]。

2.2.2 HKUST

香港科技大学研发的 HKUST 是由铜离子节点和苯三甲酸配体通过自组装合

成的。它的结构包括由铜离子连接的大型孔隙，这些孔隙被 BTC 配体桥接。铜离子通常形成铜氧化物立方体构型，而配体则像桥一样连接这些铜的氧化单元，形成具有三维孔隙网络的晶体结构。这种结构使 HKUST 系列 MOFs 材料拥有非常高的比表面积和可调节的孔径，适合用作气体吸附和分离材料。

HKUST-1 是著名且被广泛研究的一种金属有机框架（MOFs）材料，也是 HKUST 系列在 MOFs 中的代表，也称为 MOF-5[45]。有研究表明，HKUST-1 是由铜离子（Cu^{2+}）和 1,3,5-苯三甲酸（H_3BTC）通过配位键连接构成，具有大的孔隙结构和高比表面积。通过替换 HKUST-1 中的铜离子（Cu^{2+}）为其他金属离子（如 Zn^{2+}、Co^{2+}等），可以制备具有不同孔隙性质和化学功能的 MOFs。目前已采用溶胶-凝胶法[46]、共沉淀法[47]、电化学法[48]、微波辅助法[49]和水热法[50]等多种方法成功地合成了 HKUST-1。Chen 等[51]首次采用 HKUST-1 吸附布洛芬、茴香脑和愈创木酚制剂，药物载量分别为 0.34 g/g、0.38 g/g 和 0.40 g/g，其在表现出较高载药量的同时也显示出良好的控释性能。

Djahaniani 等[52]采用绿色环保的“一锅法”合成了改性 MOFs 复合材料，使用碱性木质素作为模拟口服递送系统的新型 pH 响应型生物聚合物载体，用布洛芬（Ibu）作为口服药物模型，研究了载药量和药物控释行为。复合材料提高了药物在低 pH 值（例如胃介质）下的稳定性，并在类似于肠道 pH 值的 6.8～7.4 范围内控制药物释放，展示了良好的 pH 控制的药物释放能力。单个取代基的掺入可以有效地调整 HKUST 系列材料的给药行为，有利于实现个性化的给药模式。活性基团的引入也可以促进合成后的修饰，以实现靶向基团的偶联。结合单一取代基的 MOFs 作为生物医学药物传递替代载体在有效药物载荷和灵活药物释放方面表现良好。

2.2.3　ZIFs

ZIFs 即沸石咪唑酯框架（zeolitic imidazolate frameworks）材料，是利用 Zn（Ⅱ）或 Co（Ⅱ）与咪唑（或咪唑衍生物）配体反应，环上面的N以四配位的方式自组装而成。合成出的类沸石结构的 MOFs 材料（容易合成）ZIFs 的孔结构比较类似于一种硅铝酸盐沸石。ZIFs 是由过渡金属离子（通常是 Zn^{2+}或 Co^{2+}）组成的沸石咪唑酸盐框架 MOFs。ZIFs 的拓扑结构与沸石同构，因此它们显示了 MOFs 和沸石的综合优点。例如，与其他一些 MOFs 相比，ZIFs 表现出较高的比表面积、较高的结晶度和灵活性，以及良好的热稳定性和化学稳定性。

ZIFs 具有一系列的结构并且易于功能化。在 ZIFs 中，常见的也是被合成最多的是 ZIF-8 与 ZIF-67，分别为 Zn 和 Co 与二甲基咪唑（2-methylimidazole）配位而成。这两种 ZIFs 合成较为简易，甚至可在室温下结晶而出。特别是 ZIF-8 具有极好的稳定性，即便在沸水、苯、甲醇中浸泡 1～7 天也能保持良好的稳定性。

在 ZIFs 家族中，ZIF-8 是非常重要的 MOFs，其具有高比表面积、无毒、生物相容性好和优良的化学和热稳定性，是最稳定的 MOFs 之一。由于 ZIF-8 的尺寸较小，这些化合物被限制在 ZIF-8 的框架内，会抑制其过早释放。ZIF-8 是高度 pH 响应的，因为它在生理条件下保持稳定，但在酸性环境中易分解。且它的生物相容性较好，这依托于其所含的金属离子，此外，ZIF-8 也有抗感染的功效。董晖等设计开发了一种含 Mg^{2+}/ZIF-8 水凝胶–多孔钛合金复合体（QPMZ-Ti），Mg^{2+}、ZIF-8 和 QP 水凝胶协同用于钛合金的功能化改性，赋予钛合金促成骨和抗感染能力[53]。这种材料的生物相容性好，抗菌能力强且兼具促成骨能力，可用于改善钛合金的生物惰性。ZIF-8 还可以通过增强渗透和滞留（EPR）效应来实现肿瘤的有效摄取。

有研究用“一锅法”合成了 ZIF-8 包装挥发性和疏水的 D-*α*-生育酚琥珀酸，载药率达到了 43.03%（质量分数）。由于 ZIF-8 的 pH 敏感性，实验得到的 D-*α*-生育酚琥珀酸@ZIF-8 会在酸性环境中迅速分解，可以控制药物靶向释放[54]。迄今为止研究的 ZIF-8 的其他应用均与超疏水涂层[55]、电池[56]和药物递送[57]有关。而 ZIF-67 被用于与腐蚀相关的研究，并且已经被证明能有效地抑制腐蚀[58]。此外，它也被用于催化 [59]、电池[58]和 3D 打印 [60]等方面。

Shen 等[61]还开发了一种核壳纳米复合材料，该复合材料由 ZIF-90 组成，并通过一种基于微流体的纳米沉淀方法，用精胺修饰的缩醛右旋糖酐（SAD）包覆。这种纳米复合材料可作为多药储存库，被用于装载两种具有不同特性的药物，其中亲水性阿霉素（Dox）与 ZIF-90 框架协调连接，将疏水性光敏剂 IR780 装入 SAD 外壳，从而实现光动力学治疗与化疗的联合治疗。

有研究将晚期糖基化终末产物（AGE）抑制剂受体 4-氯-*N*-环己基-*N*-（苯基甲基）-苯甲酰胺（FPS-ZM1）负载到钴（Co）基 MOFs 中[62]。利用 ZIF-67 来制造 FPS-ZM1 封装的 ZIF-67（FZ@ZIF-67）纳米粒子（NPs）。研究结果表明，FZ@ZIF-67 NPs 对大鼠糖尿病伤口的再上皮化、胶原沉积、新血管形成和炎症的改善具有显著作用。ZIF-8/ZIF-90 仍是目前研究最受关注的 ZIF 类型 MOFs 材料（表 2.2）。

表 2.2 ZIFs 的常见配体及评价[63]

名称	比表面积/（m^2/g）	孔隙率	评价
ZIF-8	621	0.252	ZIF-8 仅含有微孔，而 ZIF-90 由微孔和中孔组成。与 ZIF-8（90 μm）相比，ZIF-90 颗粒的尺寸更小（150 nm）
ZIF-90	1119	0.571	

2.2.4 PCNs

孔–通道式框架材料（porous coordination networks，PCNs）[64]含有多个立方

八面体纳米孔笼，并在空间上形成孔笼-孔道状拓扑结构。对于 PCNs 材料而言，Cu 离子是常见的金属节点，与三羧酸基配体配位而成，例如 H_3TAB（4,4-4-S-trazine-2,4,6-triyltribenzoate）或 HTB（S-heptazine tribenzoate）。PCNs 材料中同时有孔型结构和三维正交孔道，通过小窗口连接。与 IRMOFs 相比，PCNs 结构更为复杂。Luo 等开发了具有包埋 β-雌二醇模型药物能力的低细胞毒性 Zr 基 MOF PCN-221 作为药物递送系统[65]。作为比较，他们还研究了另一种药物载体，halloysite 纳米管[66]，研究发现，与 halloysite 纳米管的 20 天释放期相比，PCN-221 中的 β-雌二醇可控制释放超过 31 天，这表明 PCN-221 是 β-雌醇的优良载体[67]。近期研究中，有研究者开发了纳米锆基卟啉金属有机框架（MOFs）PCN-222，可作为安全有效的纳米敏化剂[67]。还有实验用聚乙二醇（PEG）包被 PCN-222（PEG-PCN）负载有促氧化药物哌啶铝（PL），以实现肿瘤特异性的化学-光动力学联合治疗[68]。PL-PEG-PCN 在生物介质中表现出高的胶体稳定性，有效细胞内递送进一步提高了癌症细胞内活性氧（ROS）的水平，且在乳腺癌症细胞中可选择性使癌症细胞死亡。研究结果表明，在癌症靶向化疗-声动力学联合治疗中，前氧化剂药物负载的卟啉 MOFs 具有生物相容性，是有效的声增敏剂[69]。目前 PCN-61/PCN-66 是研究较多的 PCN 类型 MOFs 材料（表 2.3）。

表 2.3　PCNs 的常见配体及评价[70]

名称	比表面积/(m^2/g)	孔容积/(cm^3/g)	评价
PCN-61	3000	1.36	PCN-66 是 PCN-61 的延伸形式。在 PCN-61 中，配体的中心含有苯环，而在 PCN-66 中，该环被三苯胺基取代
PCN-66	4000	1.63	

此外，PCNs 也用来与金属掺杂，用于药物治疗。铂是一种贵金属，能够形成抗肿瘤药物以及形成与有机分子的金属复合物，使其能够进行身体疾病的治疗，也可用于掺杂其他金属构建新的 MOFs 材料以达到不同的治疗效果。结合关于铂类 MOFs 的研究文献，可以发现在生物医学领域，其应用大致可分为三个方面：作为生物制药分析平台、作为光催化剂和作为药物传递载体，如 Pt NPs 可与 MOFs 联合用于抗菌治疗。有研究将铂原子掺杂到卟啉金属有机框架中，形成 PCN-222-Pt，用于抗菌和抗炎，为不使用抗生素治疗牙周炎提供了一种方便、无创的新方法。

2.2.5　MILs

MILs 材料是另外一种较为特殊的 MOFs 材料，一般使用金属离子（例如 Fe、Al 和 Cr）与羧酸基配体（均苯三甲酸、琥珀酸、对苯二甲酸、戊二酸和等二羧酸等）配位而成；MILs 家族首先由 Ferey 和其同事合成，包括 MIL-100（Fe）[71]、MIL-101（Fe）、MIL-53（Cr、Fe、Al）[72]、Bio-MOF-Zn[73]等，已成功装载 PEG[74]、布洛芬[75]、顺铂前药[76]、尼美舒利[77]、5-氟尿嘧啶[78]、多西他赛[79]等用于给药。

MILs 材料同样具有极高的比表面积，其中 MIL-100 和 MIL-101 是最常见的两种类型。MIL-100 是由 Fe^{3+}或 Cr^{3+}与均苯三甲酸配位形成的，这种结构在 275℃下仍能保持良好的热稳定性，其比表面积最高可达到 3100 m^2/g，孔径则可达 2.9 nm。与 MIL-100 相比，MIL-101 则是过渡金属离子与对苯二甲酸配位而成，且比表面积最高可达 5900 m^2/g。这些 MOFs 具有较高的 SSA 特性，甚至超过 3000 m^2/g，并具有较大的孔径（＞2.5 nm）[80]。

研究人员经常用 Cr-MOFs 原型结构 MIL-100(Cr)和 MIL-101(Cr)来评估 MOFs 的药物载量。这两种材料是由金属八面体和二羧酸的三聚体组成，MIL-100(Cr)含有 Cr（Ⅲ）离子以及均苯三甲酸（BTC）或三聚酸，而 MIL-101(Cr)含有对苯二甲酸以及 Cr（Ⅲ）离子。一种典型的模型药物布洛芬（Ibu），装载于 Cr-MOFs 中，具有每克脱水 MIL-101(Cr)装载 1.4 g 布洛芬的能力，而 MIL-100(Cr)仅吸收 0.35 g 布洛芬[81]。

由于铬具有毒性，上述 MOFs 材料在生物医学领域的应用受到限制。因此在 MIL-53（Fe）之后不久，有研究基于对苯二甲酸酯阴离子和 Fe（Ⅲ）的八面体结构，制备了铁基 MOFs（Fe-MOFs）。由于其低毒性、结构灵活性和可生物降解性，它们被发展成为自发给药的潜在候选药物。纳米级 Fe-MOFs[66]被有效地用于装载抗癌或逆转录病毒药物；除了它们的可降解性、生物安全性和成像能力外，它们在体外和体内的作用都得到了证明。由于亚铁连接框架的多功能性，MIL-53（Fe）具有毒性低、生物相容性好的特点，可以装载抗癌药物冬凌草素，其装载潜力高达 56.25%（质量分数），缓释期超过 7 天。Luo 等将氧化石墨烯（GO）和 Pt NPs 与氨基-MIL-125 共掺杂，形成 NH_2-MIL-125-GO-Pt，用于光催化灭菌[82]。在材料的掺杂过程中，铂与 MOFs 之间存在一个肖特基结，因此在界面上有大量的电荷。氧化石墨烯优异的导电率使光生电子空穴对有效分离和转移，从而提高了光催化效率和杀死细菌的能力。MIL 类型 MOFs 材料其金属/金属簇包括 Fe、Cr、Cu 等（表 2.4）。

表 2.4　MILs 的常见配体及评价[83]

名称	比表面积/（m^2/g）	孔隙率	评价
Cu-MOF	2460	0.65	MIL-101(Cr)晶体是具有光滑表面的八面体，用于合成的不同衍生物
Fe-MOF	3020	0.82	
MIL-101(Cr)	2606	1.80	

2.2.6　UiO 系列材料

UiO（University of Oslo）材料由含 Zr（锆）的正八面体[$Zr_6O_4(OH)_4$]与 12 个对苯二甲酸有机配体相连，形成包含八面体中心孔笼和八个四面体角笼的三维微孔结构[84]。它们通过强 Zr—O 键形成一个具有高度对称性的立方体框架结构（图 2.5）。MOFs-UiO-66 由锆八面体团簇和 BDC_2 配体组成。UiO 是含 Zr 的 MOFs 中最常被报道的材料之一。大多数含 Zr 的 MOFs 具有特殊的水稳定性、理化稳定性、后改性和可生物降解性等，能够满足储气分离、水净化或催化等实际应用的要求。UiO-66、UiO-67 等系列作为载体模型，可应用于生物医学（药物递送、成像、传感器和探针等）领域。常用的 UiO 系列材料有 UiO-66[85]，UiO-67[86]、UiO-68[87]、UiO-69[88]等，其中被研究最多的是 UiO-66。

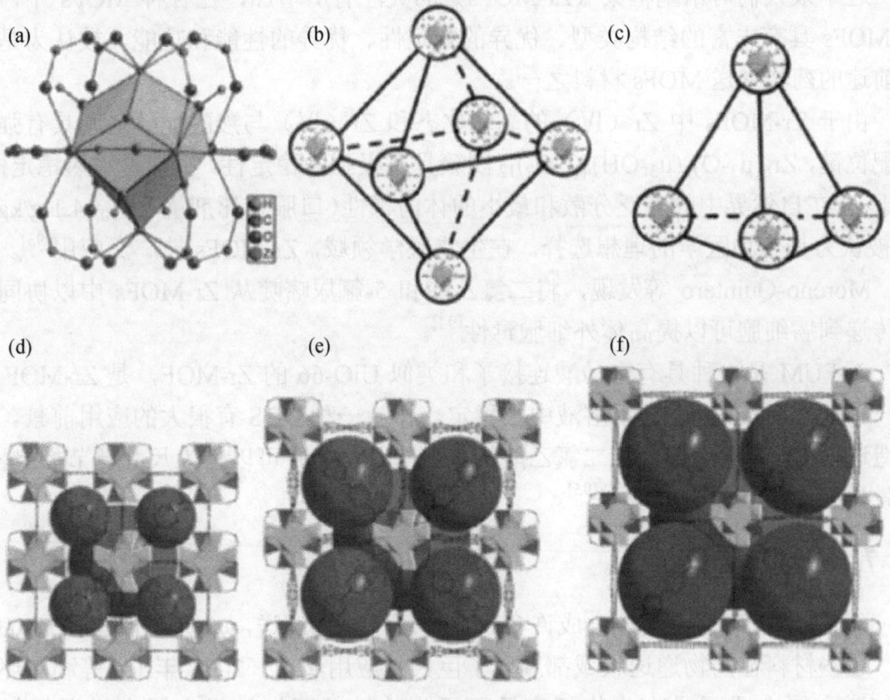

图 2.5　UiO 立方体框架结构[89]

（a）二级构筑单元；（b）八面体笼；（c）四面体笼；（d）UiO-66；（e）UiO-67；（f）UiO-68

UiO-66 的显著特点是其具有高度的热稳定性和化学稳定性，能够耐受酸、碱和水的环境。虽然多数 MOFs 材料的稳定性较差，但 UiO-66 材料却具有较好的热稳定性，其结构可在 500℃保持稳定，而且它还具有很高的耐酸性和一定的耐碱性（表 2.5）。

表 2.5　UIO 的常见配体及评价[90, 91]

名称	比表面积/（m^2/g）	孔隙率	评价
UiO-66-PDC	1081.6	—	电化学合成的 MOFs 比通过水热和微波方法制备的 MOFs 具有更高的表面积。
UiO-66-NH_2	863	0.524	
UiO-66-TLA	437	0.529	与 UiO-66-NH_2 相比，加入三甲基丙烯酸后 UiO-66-TLA 的表面积显著降低

Liu 等采用多功能胺化 UiO-67 金属有机框架设计并构建了一种新的多功能药物载体 5-Fu/UiO-67-$(NH)_2$-FAM/PMT（FUFP），纳米载体为 5-Fu/UiO-67-$(NH)_2$，5-羧基荧光素（5-FAM）作为荧光成像剂，培美曲塞（Pmt）作为叶酸（FA）靶点和抗肿瘤药物。该药物载体具有多种功能，包括共载药物、靶向传递和荧光成像，以增强肿瘤治疗的抗肿瘤细胞毒性[92]。

近年来人们对锆基框架（Zr-MOFs）的关注有所增加，在各种 MOFs 中，锆基 MOFs 具有丰富的结构类型、优异的稳定性、优异的性能和功能，被认为是最有前途的药物递送 MOFs 材料之一。

由于 Zr-MOFs 中 Zr（Ⅳ）的高氧化态和 Zr（Ⅳ）与羧酸配体之间具有强大的配位键，$Zr_6(\mu_3\text{-}O)_4(\mu_3\text{-}OH)_4(CO_2)_{12}$ 团簇具有极强的稳定性，特别是水热稳定性。由于 Zr 在自然界中的广泛分散和最小的体内毒性（口服致死剂量 LD_{50}=4.1 g/kg），Zr 被认为是生物医学的理想选择。在生物医学领域，Zr-MOFs 被广泛应用[93]。另外，Moreno-Quintero 等发现，将二氯乙酸和 5-氟尿嘧啶从 Zr-MOFs 中以协同方式传递到癌细胞可以提高体外细胞毒性[94]。

Zr-FUM 是一种具有富马酸连接子和类似 UiO-66 的 Zr-MOF，是 Zr-MOF 的另一种类型。Zr-FUM 在水溶液中很稳定，作为一种 DDS 有很大的应用前景。在制造过程中，将抗癌化合物二氯乙酸加入到 Zr-FUM 中可以作为尺寸调节控制器，负载量为 20%（质量分数）[95]。

2.2.7　其他 MOFs

除此之外，研究人员还合成许多 MOFs 用于药物递送，如 CD-MOFs、Ti-MOFs 等，这些材料在药物递送领域都展现了巨大的应用潜力。2015 年 Lu 等利用环糊精与碱金属合成了具有生物相容及可再生特性的环糊精类金属有机框架材料（CD-MOFs）[96]，为分子构筑材料家族增添了一位新的成员。环糊精和钾离子构成的环糊精金属有机框架组成成分安全无毒，是一种可靠、对称和渗透的具有超高表面积的 MOFs 材料。它所具有的良好的生物相容性，可以解决 MOFs 在药物递送时实际应用的不足，还可以有效改善客体药物分子的稳定风险，提高疗效。因此，该材料是一种极具潜力的药物载体。此外，环糊精具有的独一无二的内疏水、外亲水型空腔结构使其在医药领域得到了广泛的应用。例如，CD-MOFs 主要作为口服给药载体使用。可有效提高布洛芬、兰索拉唑、阿齐沙坦等药物的稳定性、生物利用度和半衰期。

Singh 等[97]应用简单的后修饰方法获得了生物功能化的γ-CD-MOFs，以通过透明质酸（HA）改善细胞相互作用。γ-CD-Fu 与 HA 的结合（γ-CD-Fu-HA）显示出比γ-CD-Fu 高 4.8%的平均载药能力。研究发现，γ-CD-MOFs-HA 表现出 pH 响应性药物释放，并增强 HeLa 细胞中阿霉素（Dox）的细胞吸收。因此，这个热点获得了研究者们的广泛关注并快速扩张。在以往的研究中，药物通过浸渍、研磨和共结晶被有效地纳入 CD-MOFs 中[98]。

有研究将阿齐沙坦（Azl）装载入 CD-MOFs 中，使 Azl-SD 大鼠的 Azl 生物利用度提高了 9.7 倍[99]。此外，与纯药物相比，Azl/CD-MOFs 的表观溶解度增加了 340 倍。截止到目前，CD-MOFs 已被报道用于口服、静脉注射，甚至肺部的药物递送。

CD-MOFs 还可以作为非甾体抗炎药的口服给药系统，应用于骨关节炎，或通过交联转化成立方形凝胶（由原始 γ-CD-MOFs 结构模板）用于缓释系统，如在动物牙周炎模型中，使用这些凝胶输送碘[100]。

Li 等[101]通过环糊精修饰磁性纳米粒子，制备出一种新型的核壳结构功能复合物，该复合物可用作药物载体，实现靶器官或靶组织中的控制释放。该复合物克服了磁性纳米粒子生物利用度低、载药量低的缺点，并且利用环糊精的疏水性空腔提高了药物的生物利用度。此外，通过与特异性靶向分子结合，可以实现双重靶向效果，大大改善了磁靶向载药系统的性能。

含有 Ti 金属的金属有机框架（Ti-MOFs）是一种网状复合物。这种特殊的结构是由具有羧基或含固氮杂环化合物等特定官能团的钛基阳离子或金属团簇的自组装反应引起的。Ti-MOFs 具有较大的比表面积、高孔隙率和可调的孔隙尺寸等。此外，与其他 MOFs 相比，Ti-MOFs 具有良好的生物相容性、良好的催化氧化性能和光催化性能等独特的优点[102]。目前的研究表明，所有 Ti-MOFs 都具有良好的生物相容性，这也表明其在生物医学领域具有良好的应用前景。

2018 年，为了开发 pH 响应性抗菌策略，MOF@Levo/LBL 通过阴极电泳沉积将纳米颗粒加载到胶原修饰的 Ti 基质上制备，释放羟基阴离子，与细胞外环境的持续相互作用导致细菌 ATP 水平的降低[103]。此外，钛基 MOFs 在其他治疗领域也发挥了重要作用，研究人员目前正在探索其在口服药物治疗和骨损伤中的应用。

除了 Bio-MOFs 和 CD-MOFs，MOF-74 系列也可能是潜在的药物载体。MOF-74 系列材料不仅具有无毒的 Mg^{2+}或 Zn^{2+}阳离子，而且具有非常大的孔体积[104]，这为开发可靠的药物递送系统提供了良好的基础。

2.3　常用的 MOFs 制备方法

合成 MOFs 的实验条件会影响其形貌、孔隙大小和结晶度，并能决定所得产物的理化性质。因此，正确地选择合成方法是非常重要的。此外，还应考虑到在大规模综合应用中特别重要的经济和环境因素影响。

常用的 MOFs 合成方法有溶剂热/液热法、反微乳液法、声化学和机械化学方法、微波辅助合成等（图 2.6）。合成中需要考虑尺寸、形貌和表面功能化对 MOFs 化学性质的影响，探索最佳的合成方法，以满足药物递送应用的特殊需要。不同的合成技术有不同的优点和缺点。例如水热合成和溶剂热合成反应一般在高温高压下进行，使所得产物具有较高的热稳定性。超声波方法可以使材料均匀地成核并形成小晶体，但材料的纯度并不一致。与传统的水热/溶剂热法相比，微波加热

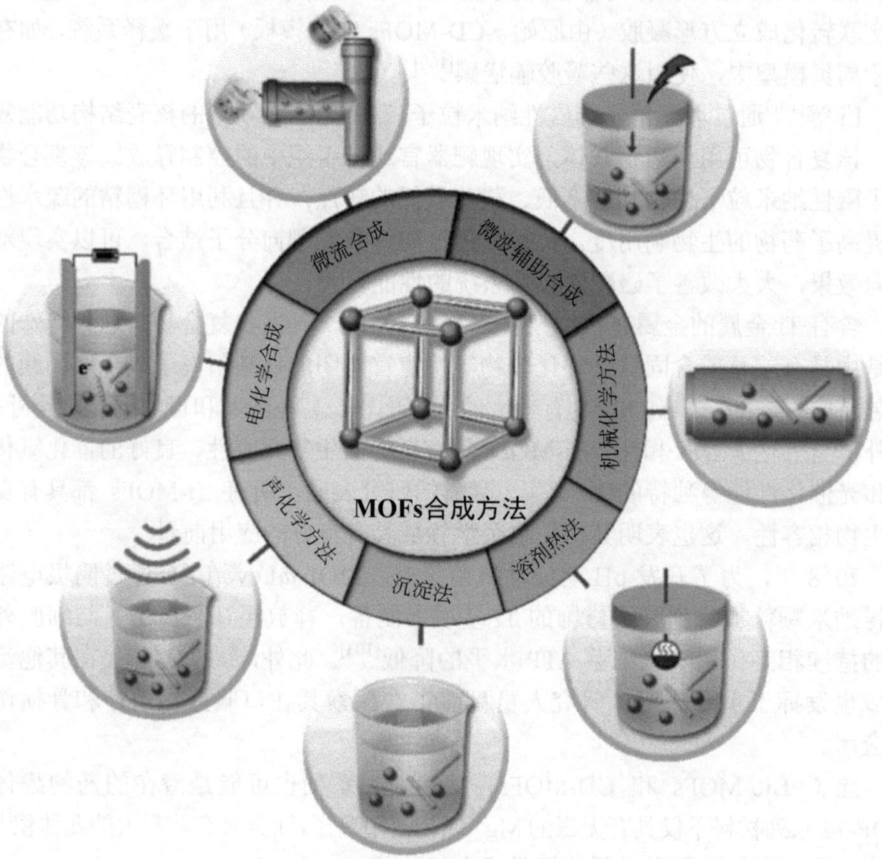

图 2.6　MOFs 常用合成方法

法大大提高了反应效率以及所制备的材料纯度。电化学合成方法不仅收率低，还容易产生副产品。因此，所得晶体的性质因不同合成条件而有所不同，说明在不同领域选择合适的合成方法至关重要。

2.3.1　水/溶剂/离子热法

自 MOFs 被发现以来，其合成方法主要是溶剂热法。典型的方法是将金属前驱体和有机连接剂溶解在溶剂中，并将其放置在封闭的反应容器中以形成自组装 MOFs 晶体。常用的溶剂包括水、*N*,*N*-二甲基甲酰胺（DMF）、*N*,*N*-二乙基甲酰胺（DEF）、甲醇、乙醇和乙腈以及它们的混合液等。合成温度通常低于 220℃，结晶时间从几小时到几十天不等。

水/溶剂热法将一定摩尔比的金属盐（离子/金属簇来源）、有机配体和溶剂/水加入反应釜中，并加热至指定温度。随着温度和反应时间的增加，反应物充分溶解混合，且釜中会产生一定的压力，金属离子和有机配体通过自组装形成相应的 MOFs 结构。当溶剂是水时，这种方法被称为水热法。通常情况下，采用水/溶剂热法制备得到的晶体尺寸较大、缺陷较少，但存在溶剂使用量大以及合成时间长等问题。因为溶剂热法中涉及一些苛刻的反应条件，如高温、高压等，故在形成 MOFs 之前，反应物前驱体通常会发生更广泛的转变，且温度和加热速率可作为额外的参数用来控制 MOFs 的成核和生长。这种方法涉及金属离子与有机连接剂的溶剂反应，并在一个封闭的容器中结晶，高温和压力（溶剂沸点以上）促进自组装和晶体生长。所选的溶剂和反应温度对反应物的溶解度都有显著的影响。应用最广泛的是有机溶剂，如丙酮、二甲基甲酰胺或乙醇。类似的方法，在水溶液中进行水热合成。通常，在溶剂热/水热过程中启动和刺激合成反应所需的能量由传统的电加热在几十个小时内提供。能量的替代形式包括电磁（微波和超声波）、电化学和机械化学能量。

Hu 等[105]通过两步溶剂热法控制结晶过程，制备了具有不同尺寸的纳米级 Mg-MOF-74 材料，后选择大小为 200 nm 的形状良好的 Mg-MOF-74 纳米棒作为进一步的药物载体。Mg-MOF-74 纳米棒对 α-氰基-4-羟基肉桂酸酯（CHC）的抗癌药的饱和负载能力约为 625 mg/g。Hu 等[106]通过三乙胺（TEA）辅助溶剂热法成功制备了直径约 200 nm 的 Zn-MOF-74 纳米棒。然后制备了双金属 Zn_xMn_{1-x}-MOF-74s，使纳米材料具有磁共振成像（MRI）能力。Gautam 等[107]研究了多孔柔性铜基金属有机框架（MOFs）Cu-BTC（也称为 HKUST-1）的合成。通过使用无毒溶剂的临时水热技术优化了晶体生长的过程。Gwon 等[108]采用水热和溶剂热反应合成了三种含有戊二酸和 1,2-双(4-吡啶基)乙烯配体的 MOF：Cu-MOF1，Co-MOF2 和 Zn-MOF3。

除了常用的有机溶剂和水，还有多种新的溶剂体系被用于制备结晶金属有机框架，比如离子液体、深共晶溶剂和表面活性剂（表 2.6）。Song 等[26]通过表面活性剂辅助的水热法合成了空心 MIL-125 纳米材料。离子液体（ILs）是一种熔点低于 100℃的离子盐，由于其独特的低蒸气特性，与传统的分子溶剂有很大的不同，其具有不可燃性、高热稳定性和高离子电导率。大多数报道的 MOFs 包含中性框架，因为金属离子的正电荷通常被有机连接剂的负电荷所平衡。然而，阳离子或阴离子可以作为构建框架的模板，形成各种离子金属有机框架，从而研究它们特定的主客体相互作用。在这种合成过程中，离子液体的阳离子和阴离子可以作为模板剂或电荷平衡中的组成部分。2002 年，在以离子液体作为反应溶剂合成金属有机配位聚合物过程中，尽管离子液体没有嵌入到配合物的结构中，但在配位聚合物的制备和结晶过程中，它起到了高效、清洁的溶剂作用。深层共晶溶剂是不同季卤化铵盐和各种氢键供体的混合物。它们也具有与离子液体相似的物理化学性质，且价格更为低廉。表面活性剂已被用作合成新的 MOFs 的反应介质。表面活性剂除了与 ILs 类似的溶剂性质外，由于其具有酸性、碱性、中性、阳离子和阴离子等多功能类型，还可以为反应体系提供更多的选择。此外，当反应温度在沸点或沸点以下时，又称非溶剂热合成。例如，沉淀过程之后是再结晶或蒸气扩散合成。蒸气扩散合成是 CD-MOFs 的第一种方法。在某些 CD-MOFs 合成中，还需要较高的反应温度才能获得优异的结晶度和较高的产率（表 2.6）。

表 2.6 溶剂热法合成的 MOFs

合成方法	中心离子	合成温度/时间	形状	参考文献
水热合成	Ni^{2+}	120℃/48 h	—	[109]
	Fe^{3+}	150℃/12 h	八面体	[110]
	Ni^{2+}, Co^{2+}	120℃/8 h	球体/薄片	[111]
	Zr^{4+}	120℃/24 h	均匀聚集的颗粒	[112]
	Zn^{2+}	140℃/24 h	菱形十二面体	[113]
	Cr^{3+}	218℃/18 h	八面体	[114]
溶剂热合成	Zn^{2+}	室温/1 h	十二面体	[115]
	Co^{2+}	室温/1 h	—	[115]
	Zr^{4+}	80 ℃/24 h	—	[115]
	Zr^{4+}	120℃/24 h	—	[116]
	Fe^{3+}	110℃/24 h	尖端的棒状	[117]
	Fe^{3+}	90～120℃/4 h	球形	[118]
	Zn^{2+}	85℃/2 d	多面体	[119]
	Cu^{2+}	120℃/24 h	不规则片状结构	[120]
	Mg^{2+}	125℃/4～24 h	六角柱	[121]

2.3.2　机械研磨法

机械研磨法是一种直接将金属盐与有机配体混合并通过机械研磨或其他手段使其充分混合的方法，然后在特定温度下进行反应以合成 MOFs。这种方法具有许多优势，包括操作简便和不需要溶剂参与，因此非常适合用于 MOFs 的大规模生产。在机械研磨法中，反应物在机械力的作用下充分混合和摩擦，使得金属离子与有机配体能够更好地接触和反应。由于不需要溶剂，机械研磨法不仅避免了溶剂的使用和回收问题，还减少了溶剂残留对产品纯度的影响。此外，机械研磨法在环保和成本方面也具有明显优势，因为其减少了溶剂的消耗和废弃物的处理需求。

该方法提供的机械能可以被用来启动机械化学合成中的反应，破坏分子内键并导致化学变化。机械化学合成方法的优点是它可以在室温下发生，并且是无溶剂的。该法温度范围大，反应时间短，而且这种方法还允许使用金属氧化物盐来代替金属盐作为前驱体，这导致的结果是产生的水将会成为反应的唯一副产物。然而由于金属氧化物基前驱体的溶解度较低，它们通常不被使用于其他合成方法。机械化学合成方法已被用于制备各种 MOFs，包括最流行的结构：MOF-5、ZIF-8、HKUST-1、MIL-101、UiO-66 等。这些 MOFs 具有高比表面积，可与传统溶剂法制备的相媲美。此外，机械化学法被成功地用于合成非常规多金属 MOFs 和以前未报道的固相。ZIFs 可通过方便、绿色无溶剂的“一锅”机械化学球磨反应在短时间内制备得到。Tanaka 等[122]通过把纳米氧化锌和 2-甲基咪唑研磨 96 小时，将氧化锌干转化为 ZIF-8，得到的 SSA 为 1480 m^2/g。

机械研磨法中，固体反应物通过研磨、过筛或剪切转化为产物，不需要过量的有机溶剂（图 2.7）。此方法具有反应时间短、环境友好等优点。Park 等[123]使用生物相容性聚合物表面活性剂对 Tb-MOF 进行简单的机械研磨，制备了在水溶液中具有良好胶体稳定性和稳定荧光性能的 Tb 基金属有机骨架纳米粒子（Tb-MOF NPs）。Tb-MOF NPs 的荧光特性使这种材料能够作为细胞成像探针。还可以利用 Tb-MOF NPs 的多孔性质，成功负载并递送抗癌药物（阿霉素）以杀死癌细胞。此外，Song 等[26]通过简单的机械研磨，制备了一种透明质酸（HA）改性的 Ti 基 MOF，系统地研究了药物递送系统在 MOF 表面上的 HA 修饰量。研究结果表明，装载有阿霉素（Dox）的纳米 MOF（MIL-125-Ti@Dox）和 HA 改性的纳米材料（MIL-125-Ti-HA@Dox）由于具有中空结构而具有较高的阿霉素装载量（25.0%～35.0%）。Quaresma 等[124]利用机械研磨方法制备了 5 种新型 Bio-MOFs，包括结合偶氮酸与内源阳离子（即 K^+、Na^+和 Mg^{2+}）：[K_2(H_2AZE)(AZE)]、[Na_4（HAZE)$_4$]、[Na_2(AZE)-(H_2O)]和多种不同的多态形式，结果表明合成的所有 MOFs 材料都具有良好的稳定性和较高的溶解度。

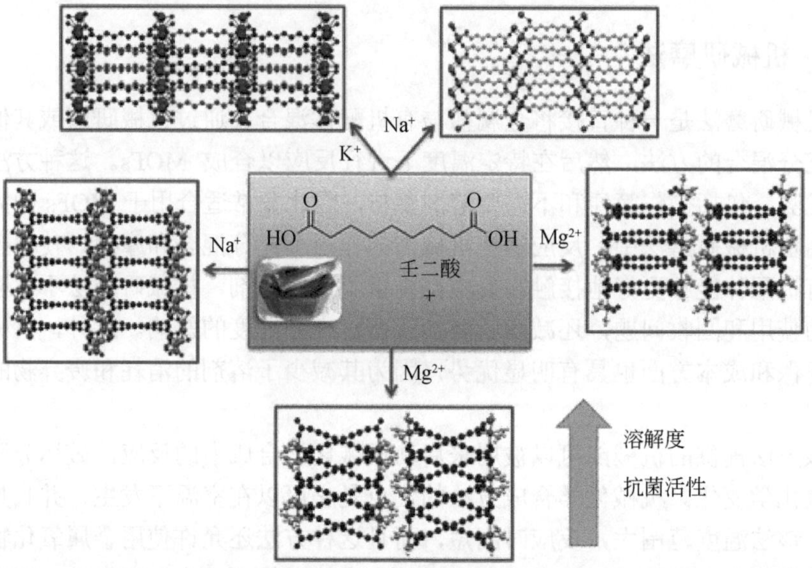

图 2.7　利用机械研磨法制备 MOF[124]

2.3.3　超声及微波辅助合成

超声法是一种将金属盐和有机配体溶解于溶剂中，通过持续超声波处理合成 MOFs 的方法。该方法具有反应条件简单、耗时短的优点，并且能够合成较小的晶体结构。然而，超声法合成的 MOFs 在纯度方面可能存在不一致的缺点。

在超声化学合成过程中，高能超声波应用于反应混合物。超声波通过液体介质传播时，会产生剧烈的空化效应，即在液体中形成和崩溃无数微小的气泡。这些气泡在崩溃时释放出局部高温高压，提供了极高的能量，促进了金属离子与有机配体的快速反应和 MOFs 的形成。只有有限数量的 MOFs 被报道可用该方法合成，如 HKUST-1、MOF-177 和 MOF-5。除了颗粒粒径，MOFs 的均匀性也受到 DMF 浓度的影响。较高的超声功率有利于相互渗透结构，而较低的超声功率会产生非穿透结构。如果中等超声强度，会产生穿透和非穿透结构互相掺杂。当液体暴露在超声波下时，液体中会产生微小气泡，这些气泡经历快速的膨胀和闭合过程。在气泡闭合的瞬间，会产生强大的冲击波，导致气泡周围形成局部高温（高达 500 K）和高压（高达 100 MPa）的环境。超声辅助合成法利用超声波的能量为溶剂中金属中心与有机配体的配位反应提供所需能量，这是一种经济、快速且环保的 MOFs 合成方法。例如，Lee 等利用超声化学合成制备了六边形的纳米材料，用于吸附水溶液中的甲基橙；Li 等利用醋酸铜和 1,3,5-苯三羧酸在超声照射下获得了直径为 100～200 nm 的苯-1,3,5-三羧酸铜（Cu-BTC）MOF。

微波辅助合成方法由于其快速成核过程，有利于在较短反应时间内制备尺寸分布均匀的 MOFs。连续流动合成具有高效连续制备具有理想尺寸和结构的单分散 MOFs 的优点。Liu 等用 100 W 微波辐照氢氧化钾、甲醇和 γ-环糊精（γ-CD）的混合物 120 分钟，然后诱导混合物结晶后洗涤，最终得到产物 γ-CD-MOFs。Guo 等通过在微波辐射的帮助下改变辐照时间，改变了 MIL-53 材料的形态和结构。Kudelin 等在微波（2.45 GHz，8 W）的辅助下合成了 Mg（2,5-二氧化物-1,4-苯二羧酸盐）（CPO-27-Mg）。实验结果表明，CPO-27-Mg 在体外能够高载药吸附阿司匹林和扑热息痛。Yao 等[125]通过微波辅助的方法合成了耐水的 MIL-101（Fe）- C_4H_4。其实验结果表明，纳米 MOFs 在 Dox 碱性水溶液中可作为纳米海绵，负载能力高达 24.5%（质量分数），而负载效率高达 98%。

Chalati 等[126]通过微波加热将等摩尔的 $FeCl_3$ 和对苯二甲酸（BDC）溶于 DMFs 溶液中，加热到 150℃，即可合成 Fe_3-(μ_3-O)Cl(H_2O)$_2$(BDC)$_3$ 的 NMOFs。扫描电子显微镜（SEM）图像显示该 NMOFs 具有不寻常的八面体形态，平均直径为 200 nm。李宗群等采用超声法合成了金属有机框架[$Zn_6(OH)_3(BTC)_3(H_2O)_3$]·$7H_2O$（Zn- MOF）和具有典型核/壳结构的 Zn-MOF@SiO_2。该工艺因其环境响应好、快速合成和高收率等优点，经常被用于制备维持微波辅助水热条件的 MOFs。微波辅助不但可以有良好的收率，而且得到的 MOFs 直径小（＜100 nm）、分散性好。MIL-100 是第一个使用微波介导的溶剂热合成的，在 220℃下连续加热 4 小时的 MOFs。微波产生的 MIL-100(Cr)具有与在相同条件下合成的框架相同的物理化学属性。目前超声辅助/微波辅助已经被广泛用于 MOFs 材料的合成（表 2.7）。

表 2.7　超声及微波辅助法合成的 MOFs

合成方法	中心离子	合成温度/时间	形状	参考文献
超声辅助合成	Cu^{2+}	130℃/24 h	八面体	[127]
	Cu^{2+}	25℃/10 min	空心，球形	[128]
	Zn^{2+}	130℃/24 h	立方体	[129]
	Eu^{3+}	40℃/21 min	均匀颗粒	[130]
	Cu^{2+}	5～60 min	球形	[131]
	Al^{3+}	150～220℃/6～24 h	非均匀结构	[132]
微波辅助合成	Al^{3+}	2 h	棱柱形	[132]
	Zn^{2+}	120℃/1 h	立方体	[133]
	Zn^{2+}	120℃/2 h	近似菱形十二面体	[134]
	Cr^{3+}	220℃/15 min 300 mW	—	[135]
	Eu^{3+}	150℃/20 min	—	[136]
	Cu^{2+}	120℃/10 min	球形	[137]

2.3.4 电化学合成

电化学合成是制备金属有机框架（MOFs）的一种有效方法，它通过在电解质溶液中施加电压或电流来促进 MOFs 的形成（表 2.8）。通常在较温和的条件下进行，不需要高温或高压环境，有利于保持有机配体的稳定性和功能完整性。通过调节电流或电压，可以快速促进金属离子与有机配体的反应，从而缩短 MOFs 形成的时间。通过精确控制电化学参数（如电流密度、电压、电解时间等），可以有效调控 MOFs 的晶体大小、形貌和孔隙性质。在某些电化学合成过程中，电解作用本身就可以提供必要的还原或氧化条件，从而减少或消除对额外化学试剂的需求。

表 2.8 电化学方法合成的 MOFs

中心离子	合成条件	形状	参考文献
Cu^{2+}	0.2～4 mA/1～48 h	八面体	[138]
Zn^{2+}	40～80 mA/1～2 h	微球	[139]
Cu^{2+}	30 V/2.5 h	八面体	[140]
Zr^{4+}	—	颗粒	[141]

金属在这一过程中提供了阳极，而连接剂分子则溶解在反应容器和电解质中。MOFs 可以通过电化学合成法无限期地制造出来。当反应混合物中包含质子溶剂时，从阳极产生的金属离子不会沉积在阴极上。根据一项研究，通过水热、非溶剂热和电化学技术得到了和 HKUST-1 MOFs 的孔隙体积和表面积几乎相同的 $Cu_3(BTC)_2$ 相。巴斯夫提供了针对 MOFs 合成的电化学技术，以避免氯离子、高氯酸盐和硝酸盐等阴离子的大规模进入。利用这种方法，在金属（包括锌、铜、镁、钴）和连接剂（1,3,5-H_3BTC、1,2,3-H_3BTC、H_2BDC 和 H_2BDC-(OH)）之间合成了多种阳极组合，这些组合产生了高孔隙率的铜、锌基 MOFs。

通常，当电流流过由浸没在固体中的金属阴极组成的电化学电池时，就形成了 MOFs，这种方法解决了有毒阴离子的问题。有研究通过电化学合成方法在室温下合成了 $Cu_3(BTC)_2$（HKUST-1）薄膜。利用铜电极作为阴极，通过在含有 1,3,5-苯三甲酸（H_3BTC）的溶液中施加电压，实现了 HKUST-1 的直接电化学生长。此外，还有研究利用电化学方法在室温条件下合成了 Fe-MIL-88B MOFs 纳米颗粒。使用铁板作为阴极，在含有有机连接器 H_2BDC（1,4-苯二甲酸）的乙醇溶液中电化学合成 Fe-MIL-88B（表 2.8）。

2.3.5　其他合成方法

1. 反相微乳法

微乳液是在表面活性剂存在下，由两种不混溶液体形成的热力学稳定的液体分散体系。这些微乳液可以看作是纳米结构的化学反应器，将 MOFs 的合成限制在纳米尺度，从而提供调整产物尺寸的可能性。合成 MOFs 之后，再开发后合成方法，使 MOFs 具备官能团。各种基团和分子能够通过共价或非共价键附着在 MOFs 上，从而支持其在生物医学领域的应用。微乳液法通常由水、油、表面活性剂和助表面活性剂组成的纳米尺度液滴作为模板来合成纳米粒子，这种方法能够有效控制 MOFs 的尺寸和形态，对于提高 MOFs 在药物递送方面的性能尤为重要。

反相微乳法基于微乳法，反相微乳液中的水滴作为反应场所，限制了纳米粒子的生长空间，有助于制备尺寸均一的纳米粒子。由于纳米粒子是在各自独立的微滴中形成的，因此可以有效避免粒子间的聚集，得到高度分散的纳米粒子。通过调整反相微乳液的组成（如水滴大小、表面活性剂种类等），可以合成不同类型和尺寸的纳米材料，包括金属、氧化物、硫化物、有机-无机杂化物等。

例如，有研究通过微乳液法合成了纳米尺度的 ZIF-8（一种基于锌和 2-甲基咪唑的 MOF），并探索了其作为抗癌药物 5-氟尿嘧啶的载体。结果表明，得到的 ZIF-8 纳米粒子展现了优异的药物装载量和缓释性能。通过室温下剧烈搅拌 ZnO_2、含顺铂的二磷酸盐与磷酸钠盐，在反相微乳液体系下反应 30 分钟，可以来制备含有顺铂前药的 MOFs 纳米颗粒。

除了合成无定形的 MOFs 纳米颗粒外，反相微乳液法也可用于合成具有晶型的 MOFs 纳米颗粒。例如，将 $GdCl_3$ 和双(甲基铵)苯-1,4-二羧酸酯按摩尔比 2∶3 在含有阳离子表面活性剂 CTAB（十六烷基三甲基溴化铵）/辛烷/1-己醇/水的微乳液体系中反应 2 小时，能够生成纳米棒。通过离心分离并用乙醇和水洗涤后，可以得到产率为 84%的纳米棒。纳米棒的形态和尺寸可以通过改变微乳液体系中的水与表面活性剂的摩尔比来调节。

2. 沉淀法、热熔挤出法、喷雾干燥法

沉淀法具有操作简便、成本低廉和易于控制的优点，适合实验室和工业规模的合成。在这种方法中，通过向溶解有反应物的溶液中加入不良溶剂，使生成的不溶于该溶剂的纳米颗粒沉淀，而反应物则保持溶解状态。该策略已成功用于合成由抗癌前药顺,反-(二氨二氯二琥珀酸盐)铂（DSCP）和 Tb 组成的 MOFs[142]。在此合成过程中，通过氢氧化钠水溶液将 $TbCl_3$ 和 DSCP 水溶液的 pH 值调至 5.5，以生成 MOFs。随后快速加入不良溶剂甲醇至反应体系中，使纳米配位聚合物

（NCPs）瞬时沉淀。透射电子显微镜（TEM）图像显示，这些 NCPs 呈球形，直径为 50～60 nm，其顺铂负载量高达 75%（质量分数），超过大多数顺铂纳米药物载体。

上述通用方法已广泛用于 MOFs 的合成。通过分别调节 MOFs 前体、反应溶剂、pH 值、温度、表面活性剂或其他模板分子，已经成功合成了一系列具有特定组成和形态的 NMOFs。研究表明，表面终止状态是决定 NMOFs 形态的重要参数。尽管 NMOFs 的合成在现象学上已得到详细描述，但对其生长机理和动力学的研究仍然稀缺。对 MOFs 生长机制和动力学过程的深入研究将促进 NMOFs 复合纳米材料在生物医学领域的进一步应用和发展。

在热熔挤出法中，挤出是一种连续加工技术。材料在挤出过程中被强制通过受限空间，并受到强烈的剪切混合作用。近年来，热熔挤出法被发现能够有效进行 MOFs 的无溶剂机械化学合成，并适用于工业批量生产。Kriesten 等以甲基纤维素为黏合剂，通过热熔挤出法制备了柔性 MOF-53 及 MOF-53-NH_2，可完全保留 MOFs 的呼吸行为[143]。

MOFs 的喷雾干燥合成是一种用于生产粉末形态的金属有机框架的工艺技术。这种方法主要是将含有 MOFs 前体的溶液通过喷嘴喷射成微小液滴，然后在热空气或惰性气体流中快速蒸发溶剂，从而形成固态的 MOFs 粒子。喷雾干燥是一种高效、可控的过程，能够在工业规模上生产具有一致性和特定粒径分布的 MOFs 粉末。通过调整喷雾参数（如喷嘴大小、流速和溶液浓度），可以精确控制粒子大小和分布。例如，Wang 等用 CD-MOFs 提高口服生物利用度并控制吲哚美辛（IMC）的释放。而 CD-MOFs 纳米晶体通过喷雾干燥技术封装，能够调节吲哚美辛的释放速率[144]。关于 MOFs 材料合成策略多种多样，而且还互相结合，多种策略混合使用（表 2.9）。

表 2.9　其他方法合成的 MOFs

合成方法	中心离子	温度/时间	形状	参考文献
	Cu^{2+}	2h	空心、球形	[145]
	Fe^{3+}	5min	空心、球形	
	Zn^{2+}	20min	空心、球形	
	Zn^{2+}	180℃	微球	[146]
喷雾干燥合成	Co^{2+}	180℃	微球	
	Zn^{2+}	180℃	微球	
	Co^{2+}			
	Zr^{4+}	150℃（90℃）	微球	[147]
	Zr^{4+}	140℃（90℃）	微球	
	Cu^{2+}	150℃	空心、球形	[148]

续表

合成方法	中心离子	温度/时间	形状	参考文献
喷雾干燥连续流合成	Cu^{2+}	100℃	空心、球形	[46]
	Zr^{4+}	180℃（115℃）	空心、球形	[149]
	Zr^{4+}	180℃（115℃）	空心、球形	[150]
	Zr^{4+}	180℃（115℃）	空心、球形	
	Zr^{4+}	180℃（115℃）	空心、球形	
	Fe^{3+}	180℃（115℃）	空心、球形	

2.4 选择机制

在药物递送领域选择合适的金属有机框架（MOFs）是一项复杂且需高度定制化的任务。选择过程需要考虑药物的性质、递送目标以及药物释放的具体要求。例如，如果目标是治疗肿瘤，并希望药物在酸性环境（如肿瘤微环境）中释放，则可以选择具有 pH 响应性的 MOFs，如那些含有可在酸性条件下解离的金属-配位键的 MOFs。选择适用于特定药物递送应用的 MOFs 需要综合考虑多个因素，包括药物的性质、所需的释放动力学、生物相容性以及经济性。通过精确控制这些参数，可以大大提高治疗效果，减少副作用，并优化患者的治疗体验。

2.4.1 金属离子的选择

金属离子的选择对于确定 MOFs 的结构、稳定性、功能性以及最终应用至关重要。金属离子的配位几何（如四面体、八面体或其他）直接影响 MOFs 的晶体结构和孔隙特性。选择具有合适配位几何的金属离子有助于形成期望的 MOFs 结构。金属离子的价态（如+2、+3 等）影响其与有机配体的配位能力和稳定性。某些价态的金属离子可能更易于形成稳定的 MOFs 结构。选择化学性质稳定的金属离子，能够抵抗合成过程中的氧化还原反应，保持 MOFs 的结构完整性。金属离子的可溶性也是一个重要因素，它影响 MOFs 合成反应的进行。选择在特定溶剂中具有良好溶解度的金属离子有利于提高反应效率和产率。

常见的用于构成 MOFs 的低毒或无毒金属离子包括 K^{+}、Ca^{2+}、Zn^{2+}、Fe^{3+}、Mg^{2+}、Cu^{2+}等。其中，Fe 基 MOFs 因其低毒性、高生物相容性被广泛应用于药物递送[151]。以 K 为金属节点的 CD-MOFs 因具有良好的载药能力及较高的生物相容性，被考虑用作肺部干粉吸入剂载体、改善难溶性药物的溶解度及生物利用度、提高药物稳定性等[152]。在 MOFs 递送抗菌药物时，可根据需要选择离子如 Cu^{2+}，Ag^{+}等[153]。MOFs 由于其结构特点，能够稳定、可持续地释放金属离子，因此被认为是金属离子储层。

MIL-88 由铁离子和 2-氨基对苯二甲酸（BDC-NH_2）配合而成，MIL-88 中的铁离子赋予了载体类酶活性，可以在肿瘤微环境中触发过量过氧化氢的 Fenton 反应生成氧。Wu 等将光敏剂多西花青（ICG）和化疗药物阿霉素（Dox）逐步封装到 MIL-88 核和 ZIF-8 壳的纳米孔中构建协同光热/光动力学/化疗纳米平台。除了有效的药物传递外，MIL-88 还可以作为一种纳米马达，将肿瘤微环境中过量的过氧化氢转化为足够的氧气，用于光动力学治疗[154]。

Peng 等设计了化疗药物通过与金属簇配合嵌入铁（III）基 MOFs 的中心[155]。合成的纳米平台可以耗尽谷胱甘肽，放大反应性氧化物种的氧化应激，并具有显著的抗癌特性。Cui 实验组[156]利用配位法合成了 Fe-MOFs 基微胶囊（图 2.8），并利用 Fe 元素的固有配位特性和 Fe^{3+}与多氧金属酸盐阴离子之间的主客体超分子相互作用合成了 DDS，这大大提高了载药/释放率（77%；83%），使其具有良好的生物相容性，且对肿瘤生长具有抑制作用。

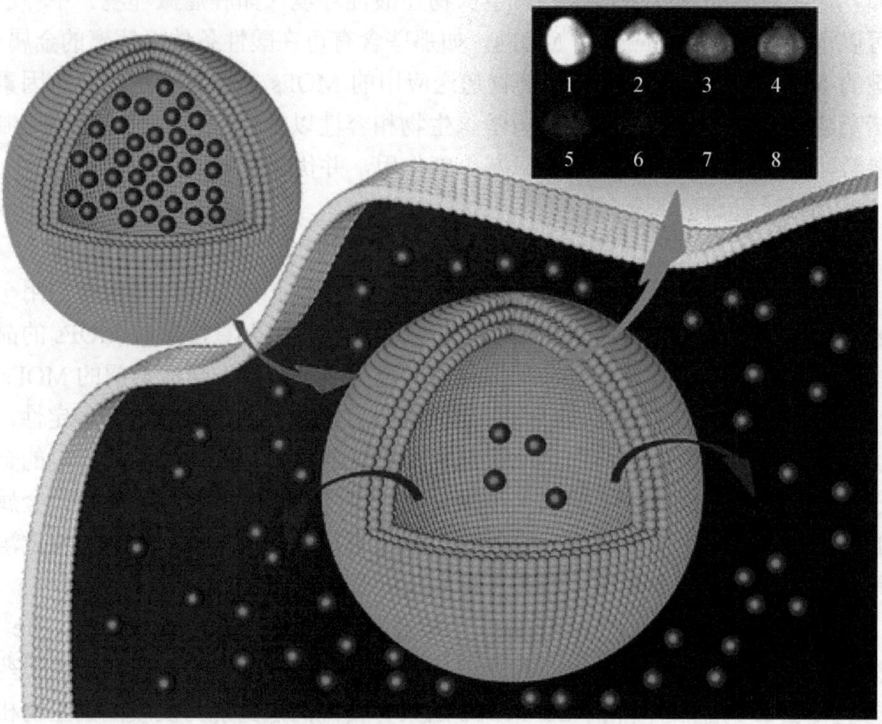

图 2.8　配位法合成 Fe-MOFs 基微胶囊[156]

含有银离子的 MOFs 材料具有抗菌的功效，银的抗菌特性和机制基本如下：

（1）带负电荷的细菌吸引带正电荷的银离子，与含硫蛋白质的细菌细胞壁作用，导致细菌结构损伤和细胞壁破裂。

（2）银纳米颗粒不仅能够增加细胞膜的通透性，而且还能穿过细胞膜，进入细胞，与细胞的内容物相互作用，改变其结构和功能。

（3）由于它们的大小和电荷不同，银纳米颗粒可以通过与不同生物成分的相互作用来影响代谢途径、细胞膜，甚至遗传物质。目前银已被证明对革兰氏阳性菌和革兰氏阴性菌都有很高的抗菌活性[157，158]。

据报道 Yang 等[159]设计了一种含石墨状碳结构的银掺杂 MOFs 衍生物（C-Zn/Ag），该衍生物具有广谱光吸收和高效的光热转换能力。C-Zn/Ag 在近红外辐照下的抑菌率为 76%，而仅释放锌离子的 C-ZIF 的抑菌率为 43%，表明协同作用提高了抗菌性能。

盐酸小檗碱（BRH）是一种异喹啉类生物碱，具有很强的生物和药理活性，但其水溶性差、半衰期短、生物利用度低[160]，导致其在临床应用中受限。李婉萌等[12]采用金属有机框架（Cu-MOFs）作为载体材料负载 BRH，以提高其生物利用度。细胞毒性实验结果显示，Cu-MOFs 材料中金属离子水平保持在生物用途允许的限度内且生物相容性良好。

Shams 等开发了一种 Cu/H_3BTC MOF161，检测了其对金黄色葡萄球菌和大肠杆菌的抑菌性能[161]。研究结果表明，其对金黄色葡萄球菌和大肠杆菌的抑制区直径分别为 22 mm 和 16 mm，抗菌活性随着 Cu/H_3BTC 浓度的增加而增加。所观察到的抑菌活性可能是 Cu^{2+}的释放引起的，从而导致细胞壁和细胞膜破裂。

此外 Huang 等将羧甲基壳聚糖（CMCS）与 HKUST-1 交织的新策略，构建了生态友好、可回收、长效、智能的 HKUST-1@CMCS 抗菌剂载体[162]。选择 Cu^{2+}为中心离子，结果表明合成的 MOFs 对金黄色葡萄球菌和大肠杆菌具有很好的抗菌活性。

除了上面提到的已经在抗菌领域被广泛研究和应用的众多金属离子外，还有其他相关的金属离子，如 Mn 和 Co 离子。虽然这些金属离子的精确抗菌过程并不能被完全解释，但它们的抗菌效果已经得到了证明。

有研究使用了可吸入颗粒大小的 γ-环糊精金属有机框架（CD-MOFs），选择活性天然化合物芍药醇（PAE）[163]作为治疗急性肺损伤（ALI）的模型药物，通过干粉吸入器实现靶向肺给药[164]。此外，姜黄素被装入环糊精基的金属有机框架（CD-MOFs）中，用于肺部药物输送[165]。与喷铣制备的微粉姜黄素相比，姜黄素负载 CD-MOFs 具有优异的空气动力学性能。

值得注意的是，ZIF-8 在酸性微环境中分解，从而诱导了装载药物的可控释放。分解产物 Zn^{2+}离子可以诱导 ROS 的产生，ROS 可以促进负载药物的内溶酶体逃逸，同时导致肿瘤细胞凋亡，增强肿瘤协同治疗效果。

通过分析可知，有机配体和金属离子的精心选择对于 MOFs 材料的成功合成和其在药物递送中的有效应用至关重要。选择合适的组成元素不仅关乎材料的结构和稳定性，还直接影响到其在生物医药领域中的安全性和功能性（表 2.10）。

表 2.10 MOFs 的各种中心离子

金属离子	MOFs
Zn^{2+}	ZIF-8
	Zn-MOF-3
Fe^{3+}	MIL-100
Zr^{2+}	UiO-66（Zr）-NO_2
	UiO-66（Zr）
Rb^{+}	CD-MOF-2（Rb）
CD^{2+}	CD-MOF-1
Hf^{2+}	Hf-MOF-808
Cu^{2+}	HKUST-1（Cu）

2.4.2 配体的选择

在探索金属有机框架（MOFs）在药物递送中的应用时，选择有机配体以控制孔隙大小和调整表面积具有重要价值。通过对羧酸盐、膦酸盐和磺酸盐配体的使用，研究人员能够设计出具有特定孔隙结构和功能的 MOFs，从而为药物的有效载药和释放提供了可能。

有机连接物在 MOFs 的三维超分子结构及其物理化学特性中起着重要的作用。最普遍的有机连接剂是羧酸盐和其他有机阴离子，如膦酸盐、磺酸盐和杂环化合物。在所有潜在的连接体中，羧酸配体约占 50%。对于应用于药物递送领域的 MOFs，所使用的连接配体除了对合成框架的物理化学性质及其生物稳定性、生物利用度[166]和毒性有显著影响以外，这些有机连接剂和官能团对 MOFs 中的药物有效载荷以及释放模式也有影响[167]。事实上，广泛的生物分子，包括氨基酸[168]、碱基[169]和糖[170]，都可以被用作构建模块。根据文献记载，2016 年，有研究通过将 Zn（Ⅱ）与一种关键的神经递质谷氨酸结合，创造了第一个基于氨基酸的生物相容性 3D MOFs。2010 年，首个基于药物的 Bio-MOFs 关于糖皮病治疗、血管舒张和抗脂质质量的研究被发表。它由内源性铁和治疗活性维生素 B3 组成。同样，一种治疗溃疡性结肠炎和其他胃肠道疾病的常用药物奥沙拉嗪，可以作为配体创建一系列新的介孔 MOFs，具有与 CPO-27/MOF-74 家族中二羟对苯二甲酸相同的配合功能。

1. 羧酸盐配体

羧酸盐配体通常可以使用与金属组分的柔性配位态来创建各种具有优异温度

稳定性的配位聚合物。因此，人们对金属羧酸盐[171]进行了大量的研究工作。在晶体结构中，通过堆积 π-π 键和氢键相互作用，可以产生一些含有芳基二羧酸盐的 MOFs。

羧酸盐配体因其与金属离子的柔性配位能力而广泛应用于 MOFs 的合成中。这种配位能力使得 MOFs 结构在经受温度变化时展现出优异的稳定性，特别是含有芳基二羧酸盐的 MOFs。在这些 MOFs 中，π-π 堆积和氢键相互作用进一步稳定了晶体结构，为药物分子提供了稳定的装载环境。这些特性使得羧酸基 MOFs 成为了药物递送中理想的载体选择，能够有效保护药物分子免受外部环境的影响，同时保持药物的生物活性。

2. 膦酸盐配体

膦酸盐配体在 MOFs 合成中的应用体现了其在调控 MOFs 孔隙结构中的重要性。通过使用线型的双膦酸或单膦酸配体，可以实现膦基团之间的有效堆积，形成柱状结构或层状结构。这种结构特性为药物分子提供了多样的装载空间和通道，有助于提高药物的载药量和释放效率。此外，非线型的膦配体也在 MOFs 合成中得到应用，进一步丰富了 MOFs 的结构多样性，为药物递送系统的设计提供了更多可能性。

线型双膦酸或单膦酸也被用作配体，它可以在磷基团之间实现足够的堆积，柱状结构-层状结构，或没有空间的层状结构。除此之外，MOFs 的合成还包括使用一些非线性膦配体双(4-羟基苯基)膦酸：用于合成具有特定孔隙结构的 MOFs，提高药物释放的选择性。三(4-羟基苯基)膦酸：构建三维网络结构的 MOFs，用于高效药物装载。

3. 磺酸盐配体

磺酸盐配体的引入为 MOFs 设计带来了新的维度。尽管磺酸基与金属离子的配位倾向相对较弱，但其配位灵活性使得磺酸盐配体能够与金属阳离子形成多样的 MOFs 结构。这种多样性为开发具有特定功能的 MOFs 提供了基础，尤其是在孔隙率和表面积调控方面。通过利用“电荷辅助氢键”等策略，磺酸盐配体 MOFs 能够形成稳定的多孔网络，为药物分子的吸附和释放提供了理想的环境。

由于磺酸盐基团的配位灵活性，磺酸盐配体可以与不同金属阳离子配位，产生不同结构的 MOFs 材料[172]。磺酸基与金属离子的相互作用与膦酸盐配体的配位倾向相比较弱；因此，利用磺酸基形成多孔固体更具挑战性。镧系金属与少量配体的反应会产生阴离子金属配体作为团簇节点。之后，由于金属配体和金属阳离子，如碱土金属的相互作用，会形成金属磺酸盐固体。“电荷辅助氢键”方法利用有机磺酸盐阴离子和多价金属阳离子建立一个持久的多孔网络，由于有机磺酸阴离子和多价金属阳离子之间形成氢键，网络中形成的六方氢键片得到了稳定。为了产生孔隙率，刚性有机磺酸阴离子将氢键片彼此分开。不同的构建孔隙方法产生了多种具有规则和永久孔隙的 MOFs。总的来说，孔隙可以允许分子在 MOFs 中的吸附和释放，富含孔道，是在给药中使用 MOFs 的基础。

4. 其他有机配体

吡啶及其衍生物配体：4,4′-联吡啶（Bipy），用于制备具有良好电子性质和光学性质的 MOFs，适用于药物的光敏递送[173]。苏州大学的研究开发了基于羟基吡啶酮和 MOFs 材料的放射性核素促排药物，成功解决了锕系元素骨骼促排的难题，在一定程度上填补了国内该领域基础研究的空白。Yang 等利用联吡啶基四羧酸配体成功构建了具有金刚石型拓扑结构的荧光 Zn(Ⅱ)MOF[174]，在 pH=5.4～6.2 的范围内，该材料显示出快速且可逆的荧光开关跃迁，能够作为 pH 发光开关，应用于 3-NPA 的荧光传感器检测中，检测限低至 1.0×10^{-6} mol/L。咪唑及其衍生物如 2-甲基咪唑（Hmim）和苯并咪唑，可以合成具有良好稳定性和生物相容性的 MOFs[175]。ZIF-8 是通过 Zn^{2+}离子与四个咪唑酸盐环[176]配位形成的，具有较高的吸附能力。有研究表明通过呋喃和三唑配体的帮助，可以提高 UiO-67 和 UiO-68 中二氧化碳的消除能力。

2.5 结构对吸附和释放药物能力的影响

金属有机框架的结构在其药物装载和释放功能中起着至关重要的作用。MOFs 的特定结构特征，包括其孔径、功能、稳定性和生物相容性，对于确定其作为药物递送系统的适用性和药物传递效率有影响，这些特性可实现高载药量、可控释放速率以及靶向体内特定部位的能力，从而显著提高治疗效果。

MOFs 中孔隙的大小直接影响其封装药物的能力。较大的孔隙可以容纳更大的药物分子或更多的药物。例如，对用单取代基官能化的 MOF-5 的研究表明，改变孔隙环境可以显著影响药物负载能力和释放行为，从而能够精确控制释放速率以获得治疗效果。具有较高表面积和孔体积的 MOFs 能够为药物相互作用和储存提供更多的空间，从而增强其载药能力。此外，有研究讨论了在超声照射下使用金属有机框架来促进抗癌药物的 pH 响应释放的过程，强调了表面积等结构参数如何促进更高的药物负载和靶向释放机制。附着在 MOFs 结构上的官能团可以提供与药物分子的特异性相互作用，例如氢键或静电相互作用，可以针对选择性药物吸附和控释进行定制。另外，还可以设计出对环境刺激（例如体内 pH 值变化）做出反应的功能化 MOFs，使药物释放更具针对性和可控性。

MOFs 结构在生理条件下的稳定性对于其作为药物递送系统的有效性至关重要。稳定的 MOFs 能够确保药物在到达靶位点之前受到保护，而其不稳定性可能导致 MOF 和药物的过早释放或降解。MOFs 的成分，特别是金属离子和有机连接剂，必须具有生物相容性，避免在医疗应用中使用时产生毒性作用。对 HKUST-1（含铜离子）和各种锆基 MOFs 等的研究表明，这类生物相容性材料可以成功在体内应用而不会产生不良反应。

MOFs 的大小和形状是影响其生物医学应用的两个重要参数。例如，可利用的纳米颗粒应小于 200 nm。为了制备单分散纳米颗粒，研究者们开发了水-溶剂热液[177]、反相微乳液法[178]和微波辅助方法[179]。除了传统的方法外，他们还引入了高通量的方法[180]，可以在短时间内更好地理解单个反应参数对产物生成的作用。Qian 等在 2018 年用三种不同的方法（水热合成、直接合成法、合成后修饰法中添加过氧化氢）合成了 MOF-5，结果表明，通过水热合成法合成的晶体具有规则的立方结构，而直接合成法和合成后修饰法中添加过氧化氢合成的晶体较大，形貌不规则。值得注意的是，对 MOFs 的尺寸探索非常重要，研究 MOFs 大小的方法有很多，而且 MOFs 材料的应用决定了它们的表征方法。如果 MOFs 纳米颗粒被用于固体基应用，那么 SEM、TEM、AFM 和 XRD 都是合适的表征方法。但是，如果在基于药物递送领域的应用中使用 MOFs，粒度分析动态光散射法测试粒径是更好的选择。在 MOFs 作为纳米载体或诊断剂或两者的生物医学应用的情况下，则需要非凝聚和胶体稳定的 MOFs。因此，液体基表征必须是澄清聚集状态。

不同配体对不同 MOFs 的药物负载和控释能力的影响不同，不同官能团对药物有效载荷和释放能力也都有影响。药物分子的释放过程可归因于三个因素：分子间的相互作用，药物分子的酸基与多孔材料的羧基之间的相互作用（如氢键），以及孔道的大小和孔的载体。分子间的相互作用容易被打破，因此药物在最初几个小时内会快速释放。通过添加官能团可以增强载体的氢键能力，从而减缓释放速率。由于配体越长，MOFs 的孔径越具有可调节性，同时比表面积是判断多孔材料吸附催化性能优劣的关键因素之一[181]。

由于 MOFs 有许多不同的结构，通过实验对药物吸附进行系统研究是不现实的，因此模拟法不仅可以避免人为因素造成的误差，还可以用来从大量样本中快速筛选出最佳候选样本，并发现实验无法提供的微观结构信息。

在传统的水溶剂热法中，一般控制参数有溶剂、反应时间、温度、化学计量学、pH 值、模板等。有研究探索了 Gd-MOFs 纳米材料作为一种潜在的磁共振成像（MRI）剂的应用。通过在反相微乳液中加入水性，可以控制 MOFs 纳米颗粒的尺寸、尺寸分布和形状[182]。MRI 结果显示（图 2.9），所有 Gd-MOFs 纳米颗粒比传统的镁具有更高的弛豫率[183]。为了加快反应过程，可以使用超声波合成法。采用超声波法可快速合成荧光微孔 MOFs、$Cu_3(BTC)_2$ 纳米晶体[131]。通过控制反应时间，使粒径在 50～100 nm；当时间增加到 30 min 时，得到了直径为 100～200 nm 的纳米晶体。随着$[Cu_2(bdc)_2(bpy)_2]_n$ 的尺寸从微米到中观尺寸晶体的减小，观察到形状记忆现象，表明尺寸对客体分子的负载和释放有影响。因此研究者需要不断调控各种参数，以获得所需 MOFs 材料。

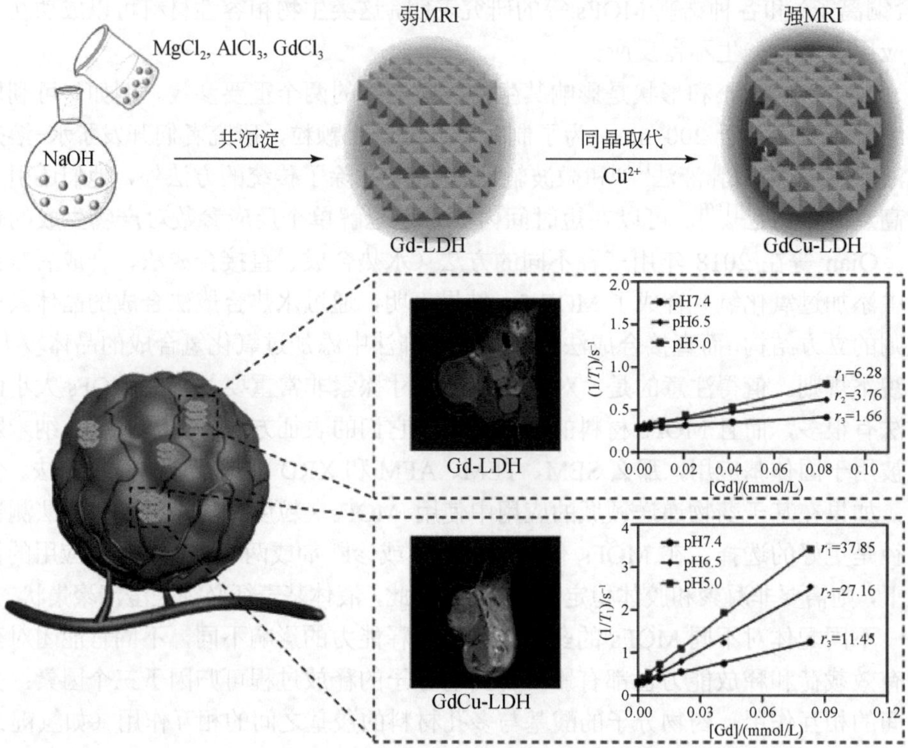

图 2.9 Gd-MOFs 在 MRI 中的使用[183]

还有研究测试了布洛芬在 MIL-100(Cr)和 MIL-101(Cr)中的吸附，显示了显著的布洛芬装载能力（0.35 g 布洛芬/g，MIL-100 和 1.4 g 布洛芬/g，MIL-101）。表明 MOFs 的金属结构对药物的吸附没有影响[184]。此外，有报道通过开发分子模拟方法，筛选了一系列 MOFs 作为模型药物氨氯地平的良好载体。对 MOF-74 系列材料中甲氨蝶呤（MTX）和 5-氟尿嘧啶的载药量的模拟研究表明，MOFs 的孔隙体积与药物吸附能力呈线性关系[185]。

MOFs 的大小也影响细胞摄取和降解，如 Orellana-Tavra 等报道，中等大小（260 nm）的 UiO-66 可以避免溶酶体降解，并将分子传递到细胞质中。然而，在溶酶体中存在小尺寸（150 nm）的 UiO-66 会进一步降解[186]，这表明中等大小的颗粒可能更适合给药。也需注意相同大小的 MOFs 可能对不同的细胞系表现出不同的细胞毒性。Cunha 等[187]嫁接不同的极性或非极性官能团到 UiO-66 来封装咖啡因和布洛芬，结果表明不同功能组有不同的药物有效载荷，而且其在修饰后可以增加对咖啡因有效装载量。

近年来，核壳 NPs 的结构得到了重大的改进。然而，探索 MOFs 在 NPs 表面的生长机制以及 MOF 壳与纳米壳核的相互作用仍存在许多挑战，但尚未见报道。基于这一想法，Huang 等[188]报道的催化剂的高性能很大程度上依赖于精心设计的具有更多活性位点和优异结构稳定性的分级中空微/纳米结构。因此，他们开发了一种简单的策略，使用核壳双 MOFs 异质结构作为前驱体和自我牺牲模板来设计和制造微笼。从双 MOFs 异质结构中衍生出金属氧化物复合材料具有挑战性，该方法可以克服不同 MOFs 之间可能的晶格失配造成的困难。一开始，药物迅速释放（38%的药物释放 220 min），然后缓慢释放（在 1000 min 内释放 17%）。与游离药物相比，该 MOFs 配方具有显著的控释效果。

在生物医学应用中也需要考虑 MOFs 的稳定性，有成千上万的 MOFs 具有不同的结构和形态，因此很难统一评价 MOFs 的稳定性。配体的酸碱度（pK_a 值）和框架的串联作用在潮湿条件下 MOFs 的稳定性或不稳定性中起着重要作用。在未互渗透的 MOFs 中，pK_a 值较高的配体在潮湿条件下更稳定。而相互渗透时，pK_a 值较低的 MOFs 比未渗透的 MOFs 更稳定。如果引入串联到连接体的组合，即使柱状连接体具有较低的碱性，MOFs 也可以是稳定的。IRMOFs-1 在潮湿的环境中相当不稳定。分子模拟结果表明，MOFs-5 和 IRMOFs-1 在含水率高达 2.3%（质量分数）时保持稳定，但晶格结构存在畸变。当含水量为 3.9%（质量分数）或更高时，由于框架中发生取代反应，框架会坍塌。例如，$Zn_4O(BDC)_3$（MOFs-5，BDC = 1,4-苯二羧酸盐）会在潮湿的空气中逐渐分解，形成无孔固体[189]。此外纳米颗粒的抗菌效果也会受其粒径和尺寸分布的显著影响[190]。金属有机框架的结构对于确定其作为药物递送系统的适用性和药物传递效率有复杂影响（表 2.11），需要在今后研究中进一步系统总结。

表 2.11　不同配体对 MOFs 载药和性能的影响

MOFs 类型	孔径/nm	载药类别	载药量/%	载药能力/%（质量分数）	参考文献
HKUST-1（Cu）	0.85～0.95	尼美舒利	—	24.20	[191]
MIL-100（Fe）	2.5～2.9	茶碱	—	32	[192]
MIL-88A（Fe）	2.4～0.2	多巴胺	—	22.70	[193]
MIL-127	0.6～1	布洛芬	—	13.60	[194]
CMC/Cu-MOF	1.1	布洛芬	93.0	279.2	[195]
Zn-GA（bio-MOF）	—	甲氨蝶呤	12.85	—	[196]
MIL-100（Fe）	2.5～2.9	多西他赛	36.4	57.2	[197]

续表

MOFs 类型	孔径/nm	载药类别	载药量/%	载药能力/%（质量分数）	参考文献
ZIF-8（Zn）	2.23	阿霉素	50	20	[198]
Fe_3O_4@MIL-100	0.6～1.6	阿霉素	37.6	10.7	[199]
Folic acid-UiO-66-NH_2 loaded-core-shell fibers	0.8～1.1	阿霉素	—	10～20	[200]
ZGGO@ZIF-8	1.16	阿霉素	93.2	—	[201]
PCN-224-DNA	1.2～1.6	阿霉素	5	—	[202]
CS/Bio-MOF	0.73	阿霉素	92.5	—	[203]
BSA/DOX@ZIF	1.16	阿霉素	19	10	[204]
IRMOF-3 @Gel	0.96	阿霉素和塞来昔布	46.85	—	[28]
PD/M-NMOF	—	阿霉素和亚甲基蓝	0.69	—	[205]
Indocyanine Green @ZIF-8	1.16	阿霉素	1.71	—	[206]
ZIF-8 @DOX@Organosilica	1.16	阿霉素	41.2	—	[57]
ZnO-DOX@ZIF-8	1.16	阿霉素	11.2	—	[207]
UCNPs@MIL-PEG	—	阿霉素	60	—	[208]
CMC/MOF-5	1.1～1.5	布洛芬	6	—	[209]

参考文献

[1] Walton K S, Snurr R Q. Applicability of the bet method for determining surface areas of microporous metal-organic frameworks. Journal of the American Chemical Society, 2007, 129: 8552-8556.

[2] Yuan Shuai, Zou Lanfang, Li Haixia, et al. Inside back cover: Flexible zirconium metal-organic frameworks as bioinspired switchable catalysts. Angewandte Chemie International Edition, 2016, 55: 10919.

[3] Klinkebiel A, Beyer O, Lüning U. Substituted 1,3,5-triazine hexacarboxylates as potential linkers for MOFs. Molecules, 2019, 24: 3480.

[4] Yang Qiang, Zhang Meiyun, Song Shunxi, et al. Surface modification of PCC filled cellulose paper by MOF-5 ($Zn_3(BDC)_2$) metal-organic frameworks for use as soft gas adsorption composite materials. cellulose, 2017, 24: 3051-3060.

[5] Tanihara Y, Nozaki A, Kuwahara Y, et al. Fabrication of densely packed HKUST-1 metal organic framework thin layers on a Cu substrate through a controlled dissolution of Cu. Bulletin of the Chemical Society of Japan, 2016, 89: 1048-1053.

[6] Chang Ruimiao, Zhang Yanyang, Zhang Guangbin, et al. Application of thermal alkaline

hydrolysis technology to improve the loading and in-vitro release of gallic acid in uio-66. Food Chemistry, 2022, 391: 133238.

[7] Mason J A, Sumida K, Herm Z R, et al. Evaluating metal-organic frameworks for post-combustion carbon dioxide capture via temperature swing adsorption. Energy & Environmental Science, 2011, 4: 3030-3040.

[8] Unlu D. Water desalination by pervaporation using MIL-101(Cr) and MIL-101(Cr)@GOdoped PVA hybrid membranes. Water, Air, & Soil Pollution, 2023, 234: 96.

[9] Cao Xiaoqiang, Wang Xuan, Chen Ming, et al. Synthesis of nanoscale zeolitic imidazolate framework-8 (ZIF-8) using reverse micro-emulsion for congo red adsorption. Separation and Purification Technology, 2021, 260: 118062.

[10] Liu Ping, Cai Kaixing, Tao Duanjian, et al. The mega-merger strategy: M@COF core-shell hybrid materials for facilitating CO_2 capture and conversion to monocyclic and polycyclic carbonates. Applied Catalysis B: Environmental, 2024, 341: 123317.

[11] Gao Yanxin, Suh M J, Kim J H, et al. Imparting multifunctionality in Zr-MOFs using the Oone-pot mixed-linker strategy: The effect of linker environment and enhanced pollutant removal. ACS Applied Materials & Interfaces, 2022, 14: 24351-24362.

[12] 李婉萌, 赵亮, 陈进, 等. 铜/腺嘌呤金属有机框架的制备并用作药物载体性能的研究. 化学世界, 2023,64: 249-256.

[13] Wang Qiuyue, Li Mingming, Sun Xinxing, et al. ZIF-8 integrated with polydopamine coating as a novel nano-platform for skin-specific drug delivery. Journal of Materials Chemistry B, 2023, 11: 1782-1797.

[14] Guo Lina, Chen Yang, Wang Ting, et al. Rational design of metal-organic frameworks to deliver methotrexate for targeted rheumatoid arthritis therapy. Journal of Controlled Release, 2021, 330: 119-131.

[15] Lin Caixue, Sun Keke, Zhang Cheng, et al. Carbon dots embedded metal organic framework@chitosan core-shell nanoparticles for vitro dual mode imaging and pH-responsive drug delivery. Microporous and Mesoporous Materials, 2020, 293: 109775.

[16] Nasrabadi M, Ghasemzadeh M A, Monfared M R Z. The preparation and charactorization of Uio-66 metal-organic frameworks for the delivery of the drug ciprofloxacin and an evaluation of their antibacterial activities. New Journal of Chemistry, 2019, 43: 16033-16040.

[17] Yadav P, Bhardwaj P, Maruthi M, et al. Metal-organic framework based drug delivery systems as smart carriers for release of poorly soluble drugs hydrochlorothiazide and dapsone. Dalton Transactions, 2023, 52: 11725-11734.

[18] Xiong Feng, Qin Zainen, Chen Haimin, et al. pH-responsive and hyaluronic acid-functionalized metal-organic frameworks for therapy of osteoarthritis. Journal of Nanobiotechnology, 2020,

18: 139.

[19] Pandey A, Kulkarni S, Vincent A P, et al. Hyaluronic acid-drug conjugate modified core-shell MOFs as pH responsive nanoplatform for multimodal therapy of Glioblastoma. International Journal of Pharmaceutics, 2020, 588: 119735.

[20] Asadollahi T, Kazemi N M, Halajian S. Alginate-zein composite modified with metal organic framework for sulfasalazine delivery. Chemical Papers, 2024, 78: 565-575.

[21] Yu Deyou, Wang Liping, Yang Taoyu, et al. Tuning lewis acidity of iron-based metal-organic frameworks for enhanced catalytic ozonation. Chemical Engineering Journal, 2021, 404: 127075.

[22] Jiang Peichun, Hu Yuling, Li Gongke. Biocompatible Au@Ag nanorod@ZIF-8 core-shell nanoparticles for surface-enhanced raman scattering imaging and drug delivery. Talanta, 2019, 200: 212-217.

[23] Ahmed A, Karami A, Sabouni R, et al. pH and ultrasound dual-responsive drug delivery system based on peg-folate-functionalized iron-based metal-organic framework for targeted doxorubicin delivery. Colloids and Surfaces A: Physicochemical and Engineering Aspects, 2021, 626: 127062.

[24] Cao Jian, Peng Xiaoxia, Li Haoyi, et al. Ultrasound-assisted continuous-flow synthesis of pegylated MIL-101(Cr) nanoparticles for hematopoietic radioprotection. Materials Science and Engineering: C, 2021, 129: 112369.

[25] Zhang Huifeng, James J, Zhao Man, et al. Improving hydrostability of ZIF-8 membranes via surface ligand exchange. Journal of Membrane Science, 2017, 532: 1-8.

[26] Song Junling, Huang Zhaoqian, Mao Jing, et al. A facile synthesis of uniform hollow MIL-125 titanium-based nanoplatform for endosomal esacpe and intracellular drug delivery. Chemical Engineering Journal, 2020, 396: 125246.

[27] Wu Wufeng, Su Jingyi, Jia Miaomiao, et al. Vapor-phase linker exchange of metal-organic frameworks. Science Advances, 2020, 6: eaax7270.

[28] Tan Guozhu, Zhong Yingtao, Yang Linlin, et al. A multifunctional MOF-based nanohybrid as injectable implant platform for drug synergistic oral cancer therapy. Chemical Engineering Journal, 2020, 390: 124446.

[29] Alves R C, Schulte Z M, Luiz M T, et al. Breast cancer targeting of a drug delivery system through postsynthetic modification of curcumin@N_3-Bio-MOF-100 via click chemistry. Inorganic Chemistry, 2021, 60: 11739-11744.

[30] Hu Xiaoxiao, Wang Caifen, Wang Lebing, et al. Nanoporous Cd-MOF particles with uniform and inhalable size for pulmonary delivery of budesonide. International Journal of Pharmaceutics, 2019, 564: 153-161.

[31] Chen Yang, Zhang Feifei, Wang Yong, et al. Recyclable ammonia uptake of a mil series of metal-organic frameworks with high structural stability. Microporous and Mesoporous Materials, 2018, 258: 170-177.

[32] Wang Jiao, Shi Mengjiao, Wang Jiajia, et al. Polycyclic polyprenylated acylphloroglucinol derivatives from hypericum acmosepalum. Molecules, 2019, 24: 50.

[33] Cabello C P, Gómez-Pozuelo G, Opanasenko M, et al. Metal-organic frameworks M-MOF-74 and M-MIL-100: Comparison of textural, acidic, and catalytic properties. ChemPlusChem, 2016, 81: 828-835.

[34] Zhang Xu, Chen An, Zhong Ming, et al. Metal-organic frameworks (MOFs) and MOF-derived materials for energy storage and conversion. Electrochemical Energy Reviews, 2019, 2: 29-104.

[35] Sun Yujia, Zhou Hongcai. Recent progress in the synthesis of metal-organic frameworks. Science and Technology of Advanced Materials, 2015, 16: 054202.

[36] Ahmad I A, Kim H, Deveci S, et al. Non-isothermal crystallisation kinetics of carbon black-graphene-based multimodal-polyethylene nanocomposites. Nanomaterials, 2019, 9: 110.

[37] Głowniak S, Szczęśniak B, Choma J, et al. Mechanochemistry: Toward green synthesis of metal-organic frameworks. Materials Today, 2021, 46: 109-124.

[38] Li Hailian, Kim J, Groy T L, et al. 20 Å Cd4in16s3514- supertetrahedral T4 clusters as building units in decorated cristobalite frameworks. Journal of the American Chemical Society, 2001, 123: 4867-4868.

[39] 徐艳红. Fe_3O_4@C@IrMOF-3 与红光碳量子点为基的多功能纳米载体设计及应用. 青岛: 青岛大学, 2019, 10: 000525.

[40] Laha D, Pal K, Chowdhuri A R, et al. Fabrication of curcumin-loaded folic acid-tagged metal organic framework for triple negative breast cancer therapy in *in vitro* and in *vivo* systems. New Journal of Chemistry, 2019, 43: 217-229.

[41] Wang Kaixin, Cai Mengru, Yin Dongge, et al. Functional metal-organic framework nanoparticles loaded with polyphyllin I for targeted tumor therapy. Journal of Science: Advanced Materials and Devices, 2023, 8: 100518.

[42] Cai Mengru, Ni Boran, Hu Xueling, et al. An investigation of IrMOF-16 as a pH-responsive drug delivery carrier of curcumin. Journal of Science: Advanced Materials and Devices, 2022, 7: 100507.

[43] Kotzabasaki M, Galdadas I, Tylianakis E, et al. Multiscale simulations reveal IrMOF-74-Iii as a potent drug carrier for gemcitabine delivery. Journal of Materials Chemistry B, 2017, 5: 3277-3282.

[44] Gong Y S, Leng J K, Guo Z M, et al. Cobalt doped in Zn-MOF-5 nanoparticles to regulate

tumor microenvironment for tumor chemo/chemodynamic therapy. Chemistry—An Asian Journal, 2022, 17: e202200392.

[45] 胡慧, 张大帅, 张秀玲, 等. 基于 HKUST-1 的新型 MOF 催化材料用于实验教学探究. 山东化工, 2020, 49: 208-210.

[46] Nuzhdin A L, Shalygin A S, Artiukha E A, et al. Hkust-1 silica aerogel composites: Novel materials for the separation of saturated and unsaturated hydrocarbons by conventional liquid chromatography. RSC Advances, 2016, 6: 62501-62507.

[47] Lidiawati N A, Nuruddin A, Nugraha. Synthesis and characterization of copper-nickel based metal organic framework by co-precipitation methode: Properties and possible application. Journal of Physics: Conference Series, 2024, 2705: 012006.

[48] Vehrenberg J, Vepsäläinen M, Macedo D S, et al. Steady-state electrochemical synthesis of Hkust-1 with polarity reversal. Microporous and Mesoporous Materials, 2020, 303: 110218.

[49] Sofi F A, Bhat M A, Majid K. Cu^{2+}-Btc based metal-organic framework: A redox accessible and redox stable MOF for selective and sensitive electrochemical sensing of acetaminophen and dopamine. New Journal of Chemistry, 2019, 43: 3119-3127.

[50] Liu Pengfei, Wen Huixiang, Jiang Zichao, et al. One-step rapid synthesis of Hkust-1 and the application for Europium(III) adsorbing in solution. Journal of Radioanalytical and Nuclear Chemistry, 2022, 331: 4309-4321.

[51] Chen Qing, Chen Qiwei, Zhuang Chong, et al. Controlled release of drug molecules in metal-organic framework material Hkust-1. Inorganic Chemistry Communications, 2017, 79: 78-81.

[52] Djahaniani H, Ghavidel N, Kazemian H. Green and facile synthesis of lignin/Hkust-1 as a novel hybrid biopolymer metal-organic-framework for a pH-controlled drug release system. International Journal of Biological Macromolecules, 2023, 242: 124627.

[53] 赵怀远. 金属有机框架衍生材料的制备及应用. 杭州: 浙江大学, 2021.

[54] Song Yuzhuo, Han Shuang, Liu Shiwei, et al. Biodegradable imprinted polymer based on ZIF-8/Dox-Ha for synergistically targeting prostate cancer cells and controlled drug release with multiple responses. ACS Applied Materials & Interfaces, 2023, 15: 25339-25353.

[55] Li Deke, Guo Zhiguang. Metal-organic framework superhydrophobic coating on kevlar fabric with efficient drag reduction and wear resistance. Applied Surface Science, 2018, 443: 548-557.

[56] Qian Ji, Li Yu, Zhang Menglu, et al. Protecting lithium/sodium metal anode with metal-organic framework based compact and robust shield. Nano Energy, 2019, 60: 866-874.

[57] Ren Shenzhen, Zhu Dan, Zhu Xiaohua, et al. Nanoscale metal-organic-frameworks coated by biodegradable organosilica for pH and redox dual responsive drug release and high-

performance anticancer therapy. ACS Applied Materials & Interfaces, 2019, 11: 20678-20688.

[58] Xu Mingqi, Wang Tong, Wang Haijun, et al. ZIF-67 on sulfur-functionalized graphene oxide for lithium-sulfur batteries. Inorganic Chemistry, 2023, 62: 3134-3140.

[59] Lu Kunkun, Song Xiufeng, Xu Lianhua, et al. Phosphorene defect/edge sites induced ultrafine COPX doping during one-pot synthesis of ZIF-67: The boosted effect on electrocatalytic oxygen reduction after carbonization. Applied Surface Science, 2019, 475: 67-74.

[60] Claessens B, Dubois N, Lefevere J, et al. 3D-printed ZIF-8 monoliths for biobutanol recovery. Industrial & Engineering Chemistry Research, 2020, 59: 8813-8824.

[61] Shen Jie, Ma Ming, Zhang Hongbo, et al. Microfluidics-assisted surface trifunctionalization of a zeolitic imidazolate framework nanocarrier for targeted and controllable multitherapies of tumors. ACS Applied Materials & Interfaces, 2020, 12: 45838-45849.

[62] Sun Yi, Bao Bingbo, Zhu Yu, et al. An FPS-ZM1-encapsulated zeolitic imidazolate framework as a dual proangiogenic drug delivery system for diabetic wound healing. Nano Research, 2022, 15: 5216-5229.

[63] Byrne C, Ristić A, Mal S, et al. Evaluation of ZIF-8 and ZIF-90 as heat storage materials by using water, methanol and ethanol as working fluids. Crystals, 2021, 11: 1422.

[64] Xu Qunna, Qiu Ruijie, Ma Jianzhong. Preparation and application of polymer-based MOFs composites. Materials Review, 2020, 34: 15153-15162.

[65] Luo Ting, Liu Hao, Liang Yong, et al. A comparison of drug delivery systems of Zr-based MOFs and halloysite nanotubes: Evaluation of β-estradiol encapsulation. ChemistrySelect, 2019, 4: 8925-8929.

[66] Xu Xiaobo, Chao Yeyan, Ma Xiaozhen, et al. A photothermally antibacterial Au@halloysite nanotubes/lignin composite hydrogel for promoting wound healing. International Journal of Biological Macromolecules, 2024, 258: 128704.

[67] Zhao Yanming, Dong Yuze, Lu Futai, et al. Coordinative integration of a metal-porphyrinic framework and TiO_2 nanoparticles for the formation of composite photocatalysts with enhanced visible-light-driven photocatalytic activities. Journal of Materials Chemistry A, 2017, 5: 15380-15389.

[68] Hoang Q T, Kim M, Kim B C, et al. Pro-oxidant drug-loaded porphyrinic zirconium metal-organic-frameworks for cancer-specific sonodynamic therapy. Colloids and Surfaces B: Biointerfaces, 2022, 209: 112189.

[69] Wen Mei, Zhao Yaoyu, Qiu Pu, et al. Efficient sonodynamic ablation of deep-seated tumors via cancer-cell-membrane camouflaged biocompatible nanosonosensitizers. Journal of Colloid and Interface Science, 2023, 644: 388-396.

[70] Pham T, Forrest K A, McDonald K, et al. Modeling PCN-61 and PCN-66: Isostructural

rht-metal-organic frameworks with distinct CO_2 sorption mechanisms. Crystal Growth & Design, 2014, 14: 5599-5607.

[71] Horcajada P, Surblé S, Serre C, et al. Synthesis and catalytic properties of MIL-100(Fe), an Iron(Ⅲ) carboxylate with large pores. Chemical Communications, 2007, 27: 2820-2822.

[72] Gordon J, Kazemian H, Rohani S. MIL-53(Fe), MIL-101, and Sba-15 porous materials: Potential platforms for drug delivery. Materials Science and Engineering: C, 2015, 47: 172-179.

[73] Armando R A M, Abuçafy M P, Graminha A E, et al. Ru-90@Bio-MOF-1: A Ruthenium(Ⅱ) metallodrug occluded in porous Zn-based MOF as a strategy to develop anticancer agents. Journal of Solid State Chemistry, 2021, 297: 122081.

[74] Su Liuhui, Wu Qiong, Tan Longfei, et al. High biocompatible ZIF-8 coated by ZrO_2 for chemo-microwave thermal tumor synergistic therapy. ACS Applied Materials & Interfaces, 2019, 11: 10520-10531.

[75] Liu Wei, Li Peisen, Zhang Hongmei. Preparation of thermosensitive microcapsules and application on the controllable drug delivery of ibuprofen. International Journal of Polymeric Materials and Polymeric Biomaterials, 2018, 67: 978-986.

[76] Lin Chunhao, Tao Yiran, Saw P E, et al. A polyprodrug-based nanoplatform for cisplatin prodrug delivery and combination cancer therapy. Chemical Communications, 2019, 55: 13987-13990.

[77] Araújo A F, Gonçalves A A, Rosado F G L, et al. Synthesis, characterization, and application of Fe-Ni bimetallic nanoparticles for the reductive degradation of nimesulide. Clean-Soil, Air, Water, 2017, 45: 1500988.

[78] Chakraborty P, Dastidar P. An easy access to topical gels of an anti-cancer prodrug (5-fluorouracil acetic acid) for self-drug-delivery applications. Chemical Communications, 2019, 55: 7683-7686.

[79] Hurry C, Villette C, Mistry H, et al. A precision dosing application for patients treated with docetaxel and G-CSF. Cancer Research, 2021, 81: 228.

[80] Zhang Hanshuo, Hu Xin, Li Tianxiao, et al. Mil series of metal organic frameworks (MOFs) as novel adsorbents for heavy metals in water: A review. Journal of Hazardous Materials, 2022, 429: 128271.

[81] Pederneira N, Aina P O, Rownaghi A A, et al. Performance of MIL-101(Cr) and MIL-101(Cr)-pore expanded as drug carriers for ibuprofen and 5-Fluorouracil delivery. ACS Applied Bio Materials, 2024, 7: 1041-1051.

[82] Li Sumei, Shan Saisai, Chen Sha, et al. Photocatalytic degradation of hazardous organic pollutants in water by Fe-MOFs and their composites: A review. Journal of Environmental

Chemical Engineering, 2021, 9: 105967.

[83] Niknam E, Panahi F, Daneshgar F, et al. Metal-organic framework MIL-101(Cr) as an efficient heterogeneous catalyst for clean synthesis of benzoazoles. ACS Omega, 2018, 3: 17135-17144.

[84] Solla E L, Micheron L, Yot P, et al. 3D reconstruction and porosity study of a hierarchical porous monolithic metal organic framework by fib-sem nanotomography. Microscopy and Microanalysis, 2016, 22: 4-5.

[85] Charles V, Yang Yong, Yuan Menglei, et al. CoO_x/UiO-66 and NiO/UiO-66 heterostructures with UiO-66 frameworks for enhanced oxygen evolution reactions. New Journal of Chemistry, 2021, 45: 14822-14830.

[86] Zhao Xiaoliang, Yu Xuezheng, Wang Xueyao, et al. Recent advances in metal-organic frameworks for the removal of heavy metal oxoanions from water. Chemical Engineering Journal, 2021, 407: 127221.

[87] Mazari S A, Hossain N, Basirun W J, et al. An overview of catalytic conversion of CO_2 into fuels and chemicals using metal organic frameworks. Process Safety and Environmental Protection, 2021, 149: 67-92.

[88] Al-Omoush M K, Polozhentsev O E, Soldatov A V. High drug loading capacity of UiO-69 metal-organic framework with linkers 2,6-naphthalenedicarboxylic acid with carboplatin. Journal of Nanoparticle Research, 2023, 26: 5.

[89] Shahini M H, Mohammadloo H E, Ramezanzadeh M, et al. Recent innovations in synthesis/characterization of advanced nano-porous metal-organic frameworks (MOFs); current/future trends with a focus on the smart anti-corrosion features. Materials Chemistry and Physics, 2022, 276: 125420.

[90] Naseri A M, Zarei M, Alizadeh S, et al. Synthesis and application of [Zr-Uio-66-Pdc-So3h]Cl MOFs to the preparation of dicyanomethylene pyridines *via* chemical and electrochemical methods. Scientific Reports, 2021, 11: 1-19.

[91] Wang Tingting, Han Lin, Li Xin, et al. Functionalized Uio-66-NH_2 by trimellitic acid for highly selective adsorption of basic blue 3 from aqueous solutions. Frontiers in Chemistry, 2022, 10: 962383.

[92] Liu Weicong, Pan Ying, Zhon Yingtaog, et al. A multifunctional aminated UiO-67 metal-organic framework for enhancing antitumor cytotoxicity through bimodal drug delivery. Chemical Engineering Journal, 2021, 412: 127899.

[93] Li Boqiong, Li Chunlin, Wang Zhenxia, et al. Preparation of Ti-Nb-Ta-Zr alloys for load-bearing biomedical applications. Rare Metals, 2019, 38: 571-576.

[94] Moreno-Quintero G, Betancur-Zapata E, Herrera-Ramírez A, et al. New hybrid scaffolds based on 5-Fu/Curcumin: Synthesis, cytotoxic, antiproliferative and pro-apoptotic effect.

Pharmaceutics, 2023, 15: 1221.
[95] Wißmann G, Schaate A, Lilienthal S, et al. Modulated synthesis of Zr-fumarate MOF. Microporous and Mesoporous Materials, 2012, 152: 64-70.
[96] Lu Hongjun, Yang Xiaoning, Li Shuxian, et al. Study on a new cyclodextrin based metal-organic framework with chiral helices. Inorganic Chemistry Communications, 2015, 61: 48-52.
[97] Singh P, Feng Jinglong, Golla V K, et al. Crosslinked and biofunctionalized γ-cyclodextrin metal organic framework to enhance cellular binding efficiency. Materials Chemistry and Physics, 2022, 289: 126496.
[98] Pan Xiaodan, Junejo S A, Tan C P, et al. Effect of potassium salts on the structure of Γ-cyclodextrin MOF and the encapsulation properties with thymol. Journal of the Science of Food and Agriculture, 2022, 102: 6387-6396.
[99] He Yuanzhi, Zhang Wei, Guo Tao, et al. Drug nanoclusters formed in confined nano-cages of Cd-MOF: Dramatic enhancement of solubility and bioavailability of azilsartan. Acta Pharmaceutica Sinica B, 2019, 9: 97-106.
[100] Volkova T, Surov A, Terekhova I. Metal-organic frameworks based on Β-Cyclodextrin: Design and selective entrapment of non-steroidal anti-inflammatory drugs. Journal of Materials Science, 2020, 55: 13193-13205.
[101] Li Ruixue, Liu Shumei, Zhao Jianqing, et al. Preparation of superparamagnetic Β-cyclodextrin-functionalized composite nanoparticles with core-shell structures. Polymer Bulletin, 2011, 66: 1125-1136.
[102] Chen Jinyi, Cheng Fan, Luo Dongwen, et al. Recent advances in Ti-Based MOFs in biomedical applications. Dalton Transactions, 2022, 51: 14817-14832.
[103] Tao Bailong, Zhao Weikang, Lin Chuanchuan, et al. Surface modification of titanium implants by ZIF-8@Levo/Lbl coating for inhibition of bacterial-associated infection and enhancement of in vivo osseointegration. Chemical Engineering Journal, 2020, 390: 124621.
[104] Perry J J I V, Teich-McGoldrick S L, Meek S T, et al. Noble gas adsorption in metal-organic frameworks containing open metal sites. The Journal of Physical Chemistry C, 2014, 118: 11685-11698.
[105] Hu Jiaqi, Chen Yi, Zhang Hui, et al. Controlled syntheses of Mg-MOF-74 nanorods for drug delivery. Journal of Solid State Chemistry, 2021, 294: 121853.
[106] Hu Jiaqi, Chen Yi, Zhang Hui, et al. Tea-assistant synthesis of MOF-74 nanorods for drug delivery and in-vitro magnetic resonance imaging. Microporous and Mesoporous Materials, 2021, 315: 110900.
[107] Gautam S, Singhal J, Lee H K, et al. Drug delivery of paracetamol by metal-organic frameworks (HKUST-1): Improvised synthesis and investigations. Materials Today Chemistry,

2022, 23: 100647.

[108] Gwon K, Han I, Lee S, et al. Novel metal-organic framework-based photocrosslinked hydrogel system for efficient antibacterial applications. ACS Applied Materials & Interfaces, 2020, 12: 20234-20242.

[109] Wu Yufang, Liu Zewei, Peng Junjie, et al. Enhancing selective adsorption in a robust pillared-layer metal-organic framework *via* channel methylation for the recovery of C2-C3 from natural gas. ACS Applied Materials & Interfaces, 2020, 12: 51499-51505.

[110] Simon M A, Anggraeni E, Soetaredjo F E, et al. Hydrothermal synthesize of Hf-free MIL-100(Fe) for isoniazid-drug delivery. Scientific Reports, 2019, 9: 16907.

[111] Sun Shuyang, Huang Minjun, Wang Pengcheng, et al. Controllable hydrothermal synthesis of Ni/Co MOF as hybrid advanced electrode materials for supercapacitor. Journal of the Electrochemical Society, 2019, 166: A1799.

[112] Tambat S N, Sane P K, Suresh S, et al. Hydrothermal synthesis of NH_2-UiO-66 and its application for adsorptive removal of dye. Advanced Powder Technology, 2018, 29: 2626-2632.

[113] Butova V V, Budnyk A P, Bulanova E A, et al. Hydrothermal synthesis of high surface area Zif-8 with minimal use of tea. Solid State Sciences, 2017, 69: 13-21.

[114] Bromberg L, Diao Ying, Wu Huimeng, et al. Chromium(III) terephthalate metal organic framework (MIL-101): Hf-free synthesis, structure, polyoxometalate composites, and catalytic properties. Chemistry of Materials, 2012, 24: 1664-1675.

[115] Yoon S, Calvo J J, So M C. Removal of acid orange 7 from aqueous solution by metal-organic frameworks. Crystals, 2018, 9: 17.

[116] Lee J, Ka D, Jung H, et al. UiO-66-NH_2 and zeolite-templated carbon composites for the degradation and adsorption of nerve agents. Molecules, 2021, 26: 3837.

[117] Kim S N, Park C G, Huh B K, et al. Metal-organic frameworks, NH_2-MIL-88(Fe), as carriers for ophthalmic delivery of brimonidine. Acta Biomaterialia, 2018, 79: 344-353.

[118] Saeed T, Naeem A, Din I U, et al. Synthesis of chitosan composite of metal-organic framework for the adsorption of dyes; kinetic and thermodynamic approach. Journal of Hazardous Materials, 2022, 427: 127902.

[119] Guo Yuanyuan, Dong Anrui, Huang Qi, et al. Hierarchical N-doped cnts grafted onto MOF-derived porous carbon nanomaterials for efficient oxygen reduction. Journal of Colloid and Interface Science, 2022, 606: 1833-1841.

[120] Arul P, Huang Shengtung, Gowthaman N S K, et al. Surfactant-free solvothermal synthesis of Cu-MOF via protonation-deprotonation approach: A morphological dependent electrocatalytic activity for therapeutic drugs. Microchimica Acta, 2020, 187: 650.

[121] Lee C T, Shin M W. Solvothermal growth of Mg-MOF-74 films on carboxylic functionalized silicon substrate using acrylic acid. Surfaces and Interfaces, 2021, 22: 100845.

[122] Tanaka H, Ohsaki S, Hiraide S, et al. Adsorption-induced structural transition of ZIF-8: A combined experimental and simulation study. The Journal of Physical Chemistry C, 2014, 118: 8445-8454.

[123] Park K M, Kim H, Murray J, et al. A Facile preparation method for nanosized MOFs as a multifunctional material for cellular imaging and drug delivery. Supramolecular Chemistry, 2017, 29: 441-445.

[124] Quaresma S, André V, Antunes A M M, et al. Novel antibacterial azelaic acid BioMOFs. Crystal Growth & Design, 2020, 20: 370-382.

[125] Yao Lijia, Tang Ying, Cao Wenqian, et al. Highly efficient encapsulation of doxorubicin hydrochloride in metal-organic frameworks for synergistic chemotherapy and chemodynamic therapy. ACS Biomaterials Science & Engineering, 2021, 7: 4999-5006.

[126] Chalati T, Horcajada P, Couvreur P, et al. Porous metal organic framework nanoparticles to address the challenges related to busulfan encapsulation. Nanomedicine, 2011, 6: 1683-1695.

[127] Azad F N, Ghaedi M, Dashtian K, et al. Ultrasonically assisted hydrothermal synthesis of activated carbon-HKUST-1-MOF hybrid for efficient simultaneous ultrasound-assisted removal of ternary organic dyes and antibacterial investigation. Ultrasonics Sonochemistry, 2016, 31: 383-393.

[128] Azizabadi O, Akbarzadeh F, Danshina S, et al. An efficient ultrasonic assisted reverse micelle synthesis route for Fe_3O_4@Cu-MOF/core-shell nanostructures and its antibacterial activities. Journal of Solid State Chemistry, 2021, 294: 121897.

[129] Askari H, Ghaedi M, Dashtian K, et al. Rapid and high-capacity ultrasonic assisted adsorption of ternary toxic anionic dyes onto MOF-5-activated carbon: Artificial neural networks, partial least squares, desirability function and isotherm and kinetic study. Ultrasonics Sonochemistry, 2017, 37: 71-82.

[130] Mirhosseini H, Shamspur T, Mostafavi A, et al. A novel ultrasonic assisted-reverse micelle procedure to synthesize Eu-MOF nanostructure with high sono/sonophotocatalytic activity: A systematic study for brilliant green dye removal. Journal of Materials Science: Materials in Electronics, 2021, 32: 22840-22859.

[131] Li Zongqun, Qiu Lingguang, Xu Tao, et al. Ultrasonic synthesis of the microporous metal-organic framework $Cu_3(BTC)_2$ at ambient temperature and pressure: An efficient and environmentally friendly method. Materials Letters, 2009, 63: 78-80.

[132] Al-Attri R, Halladj R, Askari S. Green route of flexible Al-MOF synthesis with superior properties at low energy consumption assisted by ultrasound waves. Solid State Sciences,

2022, 123: 106782.

[133] Wang Yingming, Ge Shengsong, Cheng Wei, et al. Microwave hydrothermally synthesized metal-organic framework-5 derived C-doped ZnO with enhanced photocatalytic degradation of rhodamine B. Langmuir, 2020, 36: 9658-9667.

[134] Sun Li, Shao Qian, Zhang Yu, et al. N self-doped ZnO derived from microwave hydrothermal synthesized zeolitic imidazolate framework-8 toward enhanced photocatalytic degradation of methylene blue. Journal of Colloid and Interface Science, 2020, 565: 142-155.

[135] Amaro-Gahete J, Klee R, Esquivel D, et al. Fast ultrasound-assisted synthesis of highly crystalline MIL-88a particles and their application as ethylene adsorbents. Ultrasonics Sonochemistry, 2019, 50: 59-66.

[136] Lucena M A M, Oliveira M F L, Arouca A M, et al. Application of the metal-organic framework [Eu(BTC)] as a luminescent marker for gunshot residues: A synthesis, characterization, and toxicity study. ACS Applied Materials & Interfaces, 2017, 9: 4684-4691.

[137] Liu Zhao, Ye Junwei, Rauf A, et al. A flexible fibrous membrane based on Copper(II) metal-organic framework/poly(lactic acid) composites with superior antibacterial performance. Biomaterials Science, 2021, 9: 3851-3859.

[138] Zhang Xuan, Li Yun, Goethem C V, et al. Electrochemically assisted interfacial growth of MOF membranes. Matter, 2019, 1: 1285-1292.

[139] Neto O J D L, Frós A C D O, Barros B S, et al. Rapid and efficient electrochemical synthesis of a zinc-based nano-MOF for ibuprofen adsorption. New Journal of Chemistry, 2019, 43: 5518-5524.

[140] Pirzadeh K, Ghoreysh A A, Rahimnejad M, et al. Electrochemical synthesis, characterization and application of a microstructure $Cu_3(BTC)_2$ metal organic framework for CO_2 and CH_4 separation. Korean Journal of Chemical Engineering, 2018, 35: 974-983.

[141] Wei Jinzhi, Gong Fuxin, Sun Xiaojun, et al. Rapid and low-cost electrochemical synthesis of UiO-66-NH_2 with enhanced fluorescence detection prformance. Inorganic Chemistry, 2019, 58: 6742-6747.

[142] Rieter W J, Pott Kimberly M, Taylor M L, et al. Nanoscale coordination polymers for platinum-based anticancer drug delivery. Journal of the American Chemical Society, 2008, 130: 11584-11585.

[143] Kriesten M, Schmitz J V, Siegel J, et al. Shaping of flexible metal-organic frameworks: Combining macroscopic stability and framework flexibility. European Journal of Inorganic Chemistry, 2019, 2019: 4700-4709.

[144] Wang Shanshan, Yang Xing, Lu Wangxing, et al. Spray drying encapsulation of Cd-MOF nanocrystals into eudragit® Rs microspheres for sustained drug delivery. Journal of Drug

Delivery Science and Technology, 2021, 64: 102593.

[145] Carné-Sánchez A, Imaz I, Cano-Sarabia M, et al. A spray-drying strategy for synthesis of nanoscale metal-organic frameworks and their assembly into hollow superstructures. Nature Chemistry, 2013, 5: 203-211.

[146] Chaemchuen S, Zhou Kui, Mousavi B, et al. Spray drying of zeolitic imidazolate frameworks: Investigation of crystal formation and properties. CrystEngComm, 2018, 20: 3601-3608.

[147] Avci-Camur C, Troyano J, Pérez-Carvajal J, et al. Aqueous production of spherical Zr-MOF beads via continuous-flow spray-drying. Green Chemistry, 2018, 20: 873-878.

[148] Luz I, Stewart I E, Mortensen N P, et al. Designing inhalable metal organic frameworks for pulmonary tuberculosis treatment and theragnostics via spray drying. Chemical Communications, 2020, 56: 13339-13342.

[149] Kubo M, Ishimura M, Shimada M. Improvement of production efficiency of spray-synthesized hkust-1. Advanced Powder Technology, 2021, 32: 2370-2378.

[150] Garzon-Tovar L, Cano-Sarabia M, Carne-Sanchez A, et al. A spray-drying continuous-flow method for simultaneous synthesis and shaping of microspherical high nuclearity MOF beads. Reaction Chemistry & Engineering, 2016, 1: 533-539.

[151] Nikam A N, Pandey A, Nannuri S H, et al. Hyaluronic acid-protein conjugate modified iron-based MOFs (MIL-101 (Fe)) for efficient therapy of neuroblastoma: Molecular simulation, stability and toxicity studies. Crystals, 2022, 12: 1484.

[152] Hu Xiaoxiao, Wang Caifen, Wang Lebing, et al. Nanoporous Cd-MOFparticles with uniform and inhalable size for pulmonary delivery of budesonide. International Journal of Pharmaceutics, 2019, 564: 153-161.

[153] Qi Xiaoli, Shen Ningfei, Othman A A, et al. Metal-organic framework-based nanomedicines for the treatment of intracellular bacterial infections. Pharmaceutics, 2023, 15: 1521.

[154] Wu Biyuan, Fu Jintao, Zhou Yixian, et al. Tailored core-shell dual metal-organic frameworks as a versatile nanomotor for effective synergistic antitumor therapy. Acta Pharmaceutica Sinica B, 2020, 10: 2198-2211.

[155] Ding Chong-Wei, Luo Wenzhi, Zhou Jie-Yi, et al. Hydroxo iron(III) sites in a metal-organic framework: Proton-coupled electron transfer and catalytic oxidation of alcohol with molecular oxygen. ACS Applied Materials & Interfaces, 2019, 11: 45621-45628.

[156] Cui Ruixue, Zhao Pengfe, Yan Yali, et al. Outstanding drug-loading/release capacity of hollow Fe-metal-organic framework-based microcapsules: A potential multifunctional drug-delivery platform. Inorganic Chemistry, 2021, 60: 1664-1671.

[157] Anbazhagan S, Azeez S, Morukattu G, et al. Synthesis, characterization and biological applications of mycosynthesized silver nanoparticles. 3 Biotech, 2017, 7: 333.

[158] Arenas-Vivo A, Amariei G, Aguado S, et al. An Ag-loaded photoactive nano-metal organic framework as a promising biofilm treatment. Acta Biomaterialia, 2016, 97: 490-500.

[159] Yang Ye, Wu Xizheng, He Chao, et al. Metal-organic framework/Ag-based hybrid nanoagents for rapid and synergistic bacterial eradication. ACS Applied Materials & Interfaces, 2020, 12: 13698-13708.

[160] Behl T, Singh S, Sharma N, et al. Expatiating the pharmacological and nanotechnological aspects of the alkaloidal drug berberine: Current and future trends. Molecules, 2022, 27: 3705.

[161] Shams S, Ahmad W, Memon A H, et al. Cu/H_3BTC MOF as a potential antibacterial therapeutic agent against staphylococcus aureus and escherichia Coli. New Journal of Chemistry, 2020, 44: 17671-17678.

[162] Huang Guohuan, Li Yanming, Qin Zhimei, et al. Hybridization of carboxymethyl chitosan with MOFs to construct recyclable, long-acting and intelligent antibacterial agent aarrier. Arbohydrate Polymers, 2020, 233: 115848.

[163] Li Haiyan, Zhu Jie, Wang Caifen, et al. Paeonol loaded cyclodextrin mnetal-organic framework particles for treatment of acute lung injury via inhalation. International Journal of Pharmaceutics, 2020, 578: 119649.

[164] Zhou Yong, Zhang Meijuan, Wang Caifen, et al. Solidification of volatile d-limonene by cyclodextrin metal-organic framework for pulmonary delivery via dry powder inhalers: In vitro and in vivo evaluation. International Journal of Pharmaceutics, 2021, 601: 120825.

[165] Zhou Yixian, Zhao Yiting, Niu Boyi, et al. Cyclodextrin-based metal-organic frameworks for pulmonary delivery of curcumin with improved solubility and fine aerodynamic performance. International Journal of Pharmaceutics, 2020, 588: 119777.

[166] Hidalgo T, Alonso-Nocelo M, Bouzo B L, et al. Biocompatible iron(III) carboxylate metal-organic frameworks as promising RNA nanocarriers. Nanoscale, 2020, 12: 4839-4845.

[167] Cai Mengru, Qin Liuying, You Longtai, et al. Functionalization of MOF-5 with mono-substituents: Effects on drug delivery behavior. RSC Advances, 2020, 10: 36862-36872.

[168] Subramaniyam V, Ravi P V, Pichumani M. Structure Co ordination of solitary amino acids as ligands in metal-organic frameworks (MOFs): A comprehensive review. Journal of Molecular Structure, 2022, 1251: 131931.

[169] Gao Dameng, Chen Jinghuo, Tang Sheng, et al. Simultaneous quantitative recognition of qll purines including N6-methyladenine via the host-guest interactions on a Mn-MOF. Matter, 2020, 4: 1001-1016.

[170] Lu Juan, Sun Zhuo, Zhang Xin, et al. Electrospun nanofibers modified with Ni-MOF for electrochemiluminescent determination of glucose. Microchemical Journal, 2022, 180:

107623.
[171] Kim M, Lee H S, Seo D H, et al. Melt-quenched carboxylate metal-organic framework glasses. Nature Communications, 2024, 15: 1174.
[172] Zhang Guiyang, Fei Honghan. Synthesis and applications of porous organosulfonate-based metal-organic frameworks. Topics in Current Chemistry, 2019, 377: 32.
[173] Liu Bo, Li Xiangyu, Chen Huquan, et al. Electrochemical deposition of pyrazine-templated 2d Ru@MOFs nanosheets for sensitive electrochemiluminescent immunoassay of classical swine fever virus. Microchemical Journal, 2023, 194: 109301.
[174] Yang Shuailiang, Liu Wanshan, Li Gen, et al. A pH-sensing fluorescent metal-organic framework: pH-triggered fluorescence transition and detection of mycotoxin. Inorganic Chemistry, 2020, 59: 15421-15429.
[175] Han Chaoqin, Wang Lei, Si Jincheng, et al. Reticular chemistry directed "one-pot" strategy to in situ construct organic linkers and zirconium-organic frameworks. Small, 2020, 8: 2402263.
[176] Hromadka J, Tokay B, Correia R, et al. Highly sensitive volatile organic compounds vapour measurements using a long period grating optical fibre sensor coated with metal organic framework ZIF-8. Sensors and Actuators B: Chemical, 2018, 260: 685-692.
[177] Wang Peng, Qi Xuan, Zhang Xuemin, et al. Solvent: A key in digestive ripening for monodisperse Au nanoparticles. Nanoscale Research Letters, 2017, 12: 25.
[178] Hasnain J, Jiang Y, Hou H, et al. Spontaneous emulsification induced by nanoparticle surfactants. The Journal of Chemical Physics, 2020, 153: 224705.
[179] Liang Yijun, Zhang Yu, Guo Zhirui, et al. Ultrafast preparation of monodisperse Fe_3O_4 nanoparticles by microwave-assisted thermal decomposition. Chemistry-A European Journal, 2016, 22: 11807-11815.
[180] Stock N. High-throughput investigations employing solvothermal syntheses. Microporous and Mesoporous Materials, 2010, 129: 287-295.
[181] Lan Xiong, Zhang Xiaojie, Feng Yongbao, et al. Structural engineering of metal-organic frameworks cathode materials toward high-performance flexible aqueous rechargeable Ni-Zn batteries. Materials Today Energy, 2022, 30: 101157.
[182] Rojas-Montoya I D, Fosado-Esquivel P, Henao-Holguín L V, et al. Adsorption/desorption studies of norfloxacin on brushite nanoparticles from reverse microemulsions. Adsorption, 2020, 26: 825-834.
[183] Liu Jianping, Li Li, Zhang Run, et al. The adjacent effect between Gd(III) and Cu(II) in layered double hydroxide nanoparticles synergistically enhances T1-weighted magnetic resonance imaging contrast. Nanoscale Horizons, 2023, 8: 279-290.
[184] Horcajada P, Serre C, Vallet-Regí M, et al. Metal-organic frameworks as efficient materials for

drug delivery. Angewandte Chemie International Edition, 2006, 45: 5974-5978.

[185] Erucar I, Keskin S. Computational investigation of metal organic frameworks for storage and delivery of anticancer drugs. Journal of Materials Chemistry B, 2017, 5: 7342-7351.

[186] Orellana-Tavra C, Mercado S A, Fairen-Jimenez D. Endocytosis mechanism of nano metal-organic frameworks for drug delivery. Advanced Healthcare Materials, 2016, 5: 2261-2270.

[187] Cunha D, Gaudin C, Colinet I, et al. Rationalization of the entrapping of bioactive molecules into a series of functionalized porous zirconium terephthalate MOFs. Journal of Materials Chemistry B, 2013, 1: 1101-1108.

[188] Huang Yan-Feng, Sun Xiaoyi, Huo Shuhui, et al. Core-shell dual-MOF heterostructures derived magnetic $CoFe_2O_4$/CuO (Sub)microcages with superior catalytic performance. Applied Surface Science, 2019, 466: 637-646.

[189] Kaye S S, Dailly A, Yaghi O M, et al. Impact of preparation and handling on the hydrogen storage properties of Zn_4O(1,4-Benzenedicarboxylate)$_3$ (MOF-5). Journal of the American Chemical Society, 2007, 129: 14176-14177.

[190] Sharma S, Kumar K, Thakur N, et al. The effect of shape and size of Zno nanoparticles on their antimicrobial and photocatalytic activities: A geen approach. Bulletin of Materials Science, 2019, 43: 20.

[191] Ke Fei, Yuan Yupeng, Qiu Lingguang, et al. Facile fabrication of magnetic metal-organic framework nanocomposites for potential targeted drug delivery. Journal of Materials Chemistry, 2011, 21: 3843-3848.

[192] Strzempek W, Menaszek E, Gil B. Fe-MIL-100 as drug delivery system for asthma and chronic obstructive pulmonary disease treatment and diagnosis. Microporous and Mesoporous Materials, 2019, 208: 264-270.

[193] Pinna A, Ricco R, Migheli R, et al. A MOF-based carrier for in situ dopamine delivery. RSC Advances, 2018, 8: 25664-25672.

[194] Rojas S, Colinet I, Cunha D, et al. Toward understanding drug incorporation and delivery from biocompatible metal-organic frameworks in view of cutaneous administration. ACS Omega, 2018, 3: 2994-3003.

[195] Javanbakht S, Pooresmaeil M, Namazi H. Green one-Pot synthesis of carboxymethylcellulose/Zn-based metal-organic framework/graphene oxide Bio-nanocomposite as a nanocarrier for drug delivery system. Carbohydrate Polymers, 2019, 208: 294-301.

[196] Lin Wenxin, Cui Yuanjing, Yang Yu, et al. A biocompatible metal-organic framework as a pH and temperature dual-responsive drug carrier. Dalton Transactions, 2018, 47: 15882-15887.

[197] Rezaei M, Abbasi A, Varshochian R, et al. Nanomil-100(Fe) containing docetaxel for breast cancer therapy. Artificial Cells, Nanomedicine, and Biotechnology, 2018, 46: 1390-1401.

[198] El-Bindary A A, Toson E A, Shoueir K R, et al. Metal-organic frameworks as efficient materials for drug delivery: Synthesis, characterization, antioxidant, anticancer, antibacterial and molecular docking investigation. Applied Organometallic Chemistry, 2020, 34: e5905.

[199] Bhattacharjee A, Gumma S, Purkait M K. Fe_3O_4 promoted metal organic framework MIL-100(Fe) for the controlled release of doxorubicin hydrochloride. Microporous and Mesoporous Materials, 2018, 259: 203-210.

[200] Farboudi A, Mahboobnia K, Chogan F, et al. UiO-66 metal organic framework nanoparticles loaded carboxymethyl chitosan/poly ethylene oxide/polyurethane core-shell nanofibers for controlled release of doxorubicin and folic acid. International Journal of Biological Macromolecules, 2020, 150: 178-188.

[201] Lv Ying, Ding Dandan, Zhuang Yixi, et al. Chromium-doped zinc gallogermanate@zeolitic imidazolate framework-8: A multifunctional nanoplatform for rechargeable in vivo persistent luminescence imaging and pH-responsive drug release. ACS Applied Materials & Interfaces, 2019, 11: 1907-1916.

[202] Zhang Yuanchao, Wang Qingli, Chen Guang, et al. DNA-functionalized metal-organic framework: Cell imaging targeting drug delivery and photodynamic therapy. Inorganic Chemistry, 2019, 58: 6593-6596.

[203] Abazari R, Mahjoub A R, Ataei F, et al. Chitosan immobilization on Bio-MOF nanostructures: A biocompatible pH-responsive nanocarrier for doxorubicin release on MCF-7 cell lines of human breast cancer. Inorganic Chemistry, 2018, 57: 13364-13379.

[204] Liang Zuozhong, Yang Zhiyuan, Yuan Haitao, et al. A protein@metal-organic framework nanocomposite for pH-triggered anticancer drug delivery. Dalton Transactions, 2018, 47: 10223-10228.

[205] Sharma S, Sethi K, Roy I. Magnetic nanoscale metal-organic frameworks for magnetically aided drug delivery and photodynamic therapy. New Journal of Chemistry, 2017, 41: 11860-11866.

[206] Yang Chao, Xu Jing, Yang Dandan, et al. Icg@ZIF-8: one-Step encapsulation of indocyanine green in ZIF-8 and use as a therapeutic nanoplatform. Chinese Chemical Letters, 2018, 29: 1421-1424.

[207] Zheng Cunchuan, Wang Yang, Phua S Z F, et al. Zno-Dox@ZIF-8 core-shell nanoparticles for pH-responsive drug delivery. ACS Biomaterials Science & Engineering, 2017, 3: 2223-2229.

[208] Liu Yana, Zhang Cheng, Liu Hui, et al. Controllable synthesis of up-conversion nanoparticles ucnps@MIL-PEG for pH-Responsive drug delivery and potential up-conversion luminescence/magnetic resonance dual-mode imaging. Journal of Alloys and Compounds, 2018, 749: 939-947.

[209] Javanbakht S, Pooresmaeil M, Hashemi H, et al. Carboxymethylcellulose capsulated Cu-based metal-organic framework-drug nanohybrid as a pH-sensitive nanocomposite for ibuprofen oral delivery. International Journal of Biological Macromolecules, 2018, 119: 588-596.

第3章 载体设计

3.1 载体优势

药物传递[1]是指释放生物活性物质（药物、疫苗、DNA 等）的过程，需要以一个特定的速度到一个特定的地点。第一，载体必须可控制释放，而不是爆裂释放；第二，应控制载体的降解，具有一定的生物体内稳定性，并可以对载体进行表面改性；第三，载体应高容量装载药物并有效释放。基于纳米技术的药物递送研究一直备受关注，并在临床前动物模型中显示出良好效果。各种纳米材料已被广泛研究用于药物传递，包括有机材料如聚合物、胶束、脂质体、树状大分子、环糊精，以及无机材料如石墨烯、量子点、配位聚合物、氧化铁和二氧化硅[2]等。其中，多孔材料因其独特的多孔结构，在药物装载和控释方面发挥着重要作用。与其他多孔材料相比，MOFs 具有可以调整框架的官能团、调整孔径、高表面积和易于表面修饰等优点。这些独特的特性使 MOFs 成为潜在的药物载体。

目前，用于药物递送的主要载体材料依然是各种有机高分子材料，包括天然高分子材料（如明胶、壳聚糖等）、半合成高分子化合物[3]（如醋酸纤维素酞酸酯）以及合成高分子材料（如聚乳酸、聚氨基酸等[4]）三大类。其中，天然高分子材料因其稳定性高、毒性低、成膜性好和黏度大等特点，仍然是微球的主要载体材料。然而，这些材料主要来源于壳聚糖、蛋白质、海藻盐类等天然产物，资源有限，且实际使用效果不尽理想，因此在实际应用中仍存在不少问题。而 MOFs 具有超高的孔隙率（高达 90%的自由体积）和巨大的内表面积，其比表面积非常高[5]，超过 6000 m^2/g，是优良的药物载体材料。

与沸石和多孔二氧化硅不同，有机连接剂可以调节不同分子的孔径和生物相容性，因此，基于这些特性，MOFs 激发了研究者对多孔材料科学领域的极大兴趣。例如沸石咪唑酯框架（ZIF-8）由于其高的负载能力和 pH 可控制的降解能力，在不同的地区得到了广泛的应用（图 3.1）。在抗击癌症中，能控制抗癌药物阿霉素的释放[6]。布洛芬是一种常用的抗炎药物，具有抗炎、镇痛和解热的作用，但由于其溶解性差，需要频繁给药，可能会对人体产生毒副作用，限制了其连续使用。将布洛芬吸附在特定载体上，并在体内特定部位释放，可以提高药效，从而减少药物对人体的伤害[7]。UiO-66 等 MOFs 具有高结晶度、低毒性、良好的生物相容性等优点[8]，用 MOFs 装载布洛芬可以改善其现有不足。

图 3.1 ZIF-8 负载阿霉素[6]

对于半合成和合成的高分子材料，生物相容性较差，在体内难以降解，并且一些材料对酸、碱敏感，易在体内分解，甚至可能引起一定的致敏反应，导致机体炎症和对组织、细胞的损伤。基于这些高分子材料的局限性，金属有机框架（MOFs）材料作为新型药物载体的开发和利用这一思路，已经逐渐引起国内外专家的关注和重视。MOFs 材料可以通过在金属离子和有机配体之间形成配合物来发挥不同的作用。同时，由于其较大的比表面积和大的孔隙，可作为良好的药物载体。Gao 等[9]开发了一种金属有机框架载药平台，装载甲磺酸伊马替尼。超过 95%的甲磺酸伊马替尼在前 30 分钟内被释放，这意味着这些纳米颗粒具有快速释放的能力（图 3.2）。

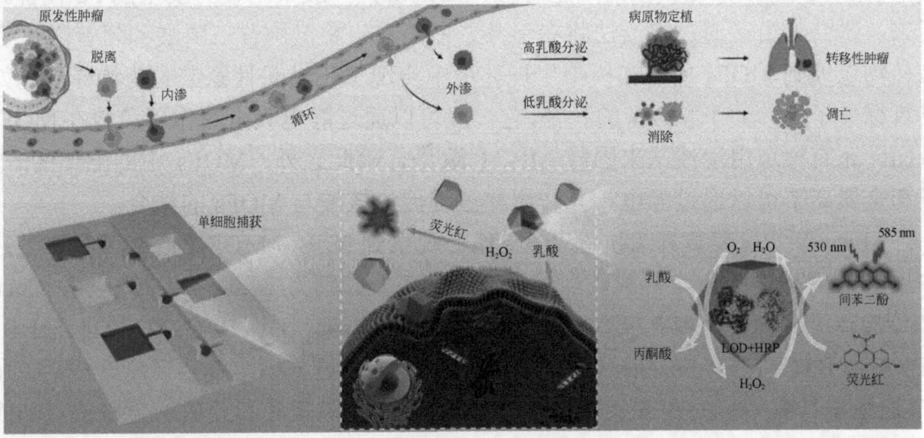

图 3.2 载酶金属有机框架辅助微流控平台实现单细胞代谢物分析[9]

在药物递送中，靶向金属有机框架（MOFs）旨在提高治疗剂递送至特定疾病部位（例如癌症治疗中的肿瘤）的精度[10]。这种靶向减少了与药物全身分布相关的副作用，并通过确保更高浓度的药物到达预期的作用部位来提高治疗效果。将药物传递到一个特定的目标。此外，它们还被用于增加对细胞的摄取和控制药物的释放。上述特征将 MOFs 转变成为理想的药物递送系统（DDS）类型之一。它们被应用于许多领域，如代谢标记分子、抗青光眼药物、激素、抗菌药物和抗癌药物。目前 MOFs 已成功装载盐酸阿霉素 [11]、甲氨蝶呤[12]、5-氟尿嘧啶[13]、咖啡因[14]、喜树碱[15]、6-巯基嘌呤[16]和庆大霉素[17]等用于靶向给药。

3.2 装 载 方 法

实现 MOFs 材料对药物的精准递送，第一步是使其有效载药。因为 MOFs 由无机金属或金属团簇节点组成，有机多齿配体充当“桥梁”，这种结构中存在一些规则和可调节的“空隙”体积（孔道、孔隙和空腔）作为可用的载体。MOFs 可以通过将药物装入“空隙”来有效地封装和交付药物。目前研究报道了浸渍法、共价键连接法、配体交换法等多种方式能够实现 MOFs 材料的有效载药。

3.2.1 浸渍法

浸渍法是将已经制备的 MOFs 材料浸入含有生物分子或药物的各种溶液中，当目标分子的尺寸小于 MOFs 的孔径时，这些分子可以进入 MOFs 的内部进行封装。例如，Horcajada 等第一次采用此法将布洛芬固载到 MIL-100 和 MIL-101 框架中[18]。由于 Cr 存在较强的细胞毒性，后期研究一般采用较为安全的 Fe 代替。后来，研究人员又通过浸渍法将布洛芬、白消安、叠氮胸苷三磷酸和阿昔洛韦[19]等药物包埋到 MOFs 内，在模拟生理环境中实现了长时间且稳定的药物释放。如果目标分子尺寸大于 MOFs 的孔径，它一般可以通过静电吸附等作用负载在 MOFs 表面。还有报道用浸渍法来固载 MP-11 酶[20]，除此之外，MOFs 中的配体的π电子和金属离子可以通过静电相互作用和氢键促进核酸与 MOFs 的结合。

Alves 等[21]以姜黄素为抗癌药物，他们观察到 N_3-Bio-MOF-100 在姜黄素溶液中浸泡 1 天和 3 天后，其载药封装效率（DLE）分别为 24.02%和 25.64%。Chen 等[22]构建了一种新型多孔镧系元素 Dy(Ⅲ)-有机框架，[Dy(BTC)(H_2O)]（H_2O）（DMF）（BTC=1,3,5- bezenetricarboxylate）体系，通过一锅溶剂热反应合成后，在 1D 通道中具有大量的开放金属位点。由于合适的孔隙尺寸和开放的金属位点功能化通道，通过简单的浸渍方法，即装载抗癌药物 5-Fu（氟尿嘧啶）。在此基础上，Wang 等开发了一种超临界溶液浸渍（SSI）技术，制备了 MOFs 的药物递送系统。首先，通过在羧基功能化的 Fe_3O_4 表面原位生长 MOFs 层，合成了核壳结构的 Fe_3O_4

@UiO-66 复合材料。该复合材料表现出超顺磁特性，允许磁性靶向和磁共振成像以及有效的药物加载和控制释放。使用 SSI 工艺和乙醇浸渍将疏水性抗癌药物姜黄素封装到合成的复合材料中，见图 3.3[23]。

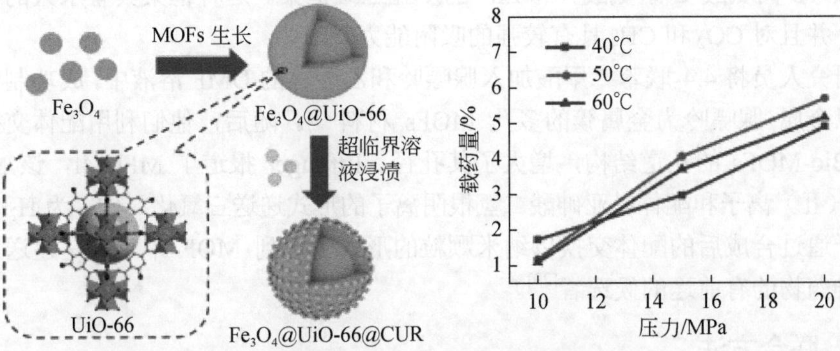

图 3.3　核壳结构的 Fe_3O_4@UiO-66 复合材料[23]

3.2.2　共价键连接法

共价键连接法是通过共价键将需要固定的药物与 MOFs 连接，实现药物固载。最常用的方法是对 MOFs 的有机配体和目标分子进行修饰或加入连接分子，利用官能团反应将目标分子和 MOFs 连接起来[24]。例如，Mirkin 等制备具有叠氮官能团结构的 UiO-66-N_3 金属有机框架材料，然后利用“点击”反应将其与带有二苄基环辛炔的 DNA 相连，把 DNA 固载在金属有机框架上。

Zimpel 等在 MIL-100（Fe）纳米颗粒的外表面上通过共价结合，以实现诸如增加化学和胶体稳定性或染料标记，基于荧光的技术对颗粒进行研究[25]。在温和条件下溶解官能化的 MOFs 后，通过液相 NMR 证明了共价纳米粒子-聚合物键的形成，MOFs 纳米颗粒与荧光标记的聚合物的功能化使得研究者能够通过荧光显微镜研究纳米颗粒是否吸收到肿瘤细胞中。此外，聚合物壳对 MIL-100（Fe）磁共振成像也有积极影响。

3.2.3　配体交换法

配体置换和金属离子置换是 Hupp 和 Farha 等总结提出的“结构基元置换”策略。该策略的特点是可以改变“框架”核心结构的组成，但通常不改变“框架”的拓扑连接[26]。这类反应通常为异相反应，框架组成的化学键需要经历断裂和重组。目前研究通常认为置换反应通常以“单晶到单晶转换”的方式发生，而不是通过“溶解-重结晶”的过程。

配体交换法将目标分子直接用作有机配体，与金属离子自组装形成一种新型

MOFs，从而提高生物相容性。例如，氨基酸具有丰富的羧基和氨基，可以用来与金属离子配位，构建 MOFs 材料[27]。Park 等利用 L-谷氨酸和 Zn^{2+}在室温下合成 Zn 基 MOFs。多肽也可以用于合成 Bio-MOFs。Rosseinsky 小组报道了一种基于二肽肌肽（β-丙氨酸-L-组氨酸）和 Zn^{2+}的多孔三维框架。这种框架具有永久的微孔结构，并且对 CO_2 和 CH_4 具有较强的吸附能力。

研究人员将 4,4-联苯二甲酸加入腺嘌呤和乙酸锌的 DMF 溶液中，成功制备了首例以金属–腺嘌呤为金属簇的多孔 MOFs 材料[28]。随后，他们利用配体交换法调控 Bio-MOFs 的孔道结构，增大了其孔径。Ettlinger 报道了 MFU-4l，该 MOF 由 Zn（Ⅱ）离子和配体以亚砷酸二氢根阴离子的形式递送三氧化二砷。为 $H_2AsO_3^-$ 阴离子通过合成后的配体交换以纳米颗粒的形式引入到 MOF 中，其是递送三氧化二砷药物的有前途的候选者[29]。

3.2.4　联合方法

1. “瓶中造船”法

在 MOFs 的制备时，采用“瓶中造船”法制备尺寸可控的金属纳米粒子。在实现封装过程中应该注意选择 MOFs 载体的孔径来匹配载药分子[30]。这种新型具有优良性能的多孔磁性核–壳复合物，结合了壳的多孔性和核的磁学性双重优势，将会在药物递送领域具有潜在的应用前景。该方法将贵重金属纳米粒子引入 MOFs 材料，将 MOFs 作为载体，通过吸附金属前体并随后还原，在 MOFs 上生成贵重金属纳米粒子。其最显著的优势是利用纳米孔道或纳米笼的限阈效应，制备尺寸可控的金属纳米粒子。实验结果发现，ZIF-8、UiO-66、γ-CD-MOFs、纳米 DUT-32、布洛芬等在合成的过程中均用到了“瓶中造船”的方法。这些药物具有较高的化学稳定性和热力学性能，有些物质可以通过改变晶体体积来增强晶体的吸附能力。

Zlotea 等以 $PdCl_2$ 作为前体，利用化学浸渍法，在 MIL-100（Al）的纳米孔道中引入了粒径为 2 nm 的 Pd 纳米粒子[31]。这种“封装策略”用于许多重要物质的合成，包括化疗药物、脱氧核糖核酸、核糖核酸或酶等。为了实现成功的封装，应选择合适的 MOFs 载体的孔径来匹配药物。一方面，微孔 MOFs 的孔径不足以输送药物，其次增加孔径会导致药物的突然释放和较低的输送能力。因此，相关研究人员采用了几种新的方法来解决这些问题。

2. 模板法

模板法是将金属盐和有机配体按一定比例溶解在含模板剂的溶液中，然后在反应釜中进行反应。此方法可以使用不同大小的模板剂来制备不同尺寸的 MOFs。

模板法最大的缺点是模板剂的去除，这可能导致部分材料的孔道塌陷，破坏孔道结构。利用一个模板分子来引导和控制 MOFs 结构的形成，这种方法通常用于生产具有特定孔隙结构和功能性的 MOFs。模板通常是一个有机分子、聚合物或其他材料，它可以在 MOFs 合成过程中起到结构定向的作用。选择合适的模板是关键步骤，模板可以是多孔的、有特定形状的或含有特定功能团的物质。这些模板在合成过程中控制了 MOFs 的孔隙大小、形状和化学功能性。在 MOFs 的合成过程中，金属离子和有机配体在模板的存在下通过自组装的方式形成 MOFs 结构。模板分子通过与金属离子或配体相互作用，指导 MOFs 的孔隙结构形成。合成完成后，通常通过加热、溶剂洗涤或其他化学处理方法将模板从 MOFs 中移除，留下具有期望孔隙结构的 MOFs。这一步是确保 MOFs 具有可用的空孔隙并实现其功能的关键。

在溶剂热条件下表面活性剂也可作为模板参与 NMOFs 的合成，这些表面活性剂分子主要包裹在 NMOFs 的表面而并不参与反应，但对 NMOFs 的形貌起重要作用。反相微乳液法的不同点在于，在加热过程中微乳液的体系会被破坏。有研究[32]报道了一种高灵敏度和高选择性的 MOFs 基材料传感器 MOF-74（Zn）-en，以 MOF-74（Zn）为模板，以乙二胺（en）为修饰基团和官能团，可以作为发光探针来测定 TBBPA（四溴双酚 A）。Liu 等通过模板诱导方法，首次诱导了两种极为罕见的由β-环糊精（β-CD）支撑的金属有机框架 CD-MOF-1 和 CD-MOF-2 的结晶。大量对照实验表明，所选模板对所得 CD-MOFs 的结晶度和孔隙率具有重要意义。具有靶向功能的 CD-MOFs 用于进行受控的药物递送和细胞毒性测定，证实了它们作为药物载体具有良好的生物学潜力[33]。

Tang 等通过结合金属有机框架（MOFs）和金属–有机薄膜的优点，提出了一种简便、高效且通用的方法来合成 pH 响应型纳米胶囊（～120 nm）[34]。ZIF-8 纳米颗粒用作模板，其上的 Fe（Ⅲ）-儿茶酚复合物薄膜涂层源自多巴胺改性藻酸盐（AlgDA）和 Fe（Ⅲ）离子之间的配位。去除模板后，得到了具有 pH 响应壁的纳米胶囊。

Zou 等[35]采用可酸降解的 MOFs 沸石咪唑酯框架 ZIF-8，既可以作为自我牺牲模板来合成尺寸和形态可控的中空介孔二氧化硅材料（HMSN），又可以作为介孔阻滞剂，用于制造基于 pH 值的 HMSN 药物递送系统（图 3.4）。

3. 生物矿化

随着 MOFs 材料纳米结构制备的发展，通过精确控制界面结构和表面化学来轻松合成大规模 MOFs 薄膜仍然具有挑战性。从布洛芬的装载和控释开始，MOFs 在药物传递中的后续应用仍主要集中在小分子药物的释放上，如抗肿瘤药物阿霉素和姜黄素等。而对于生物大分子，研究者常通过仿生物矿化实现 MOFs 封装和保护大分子。

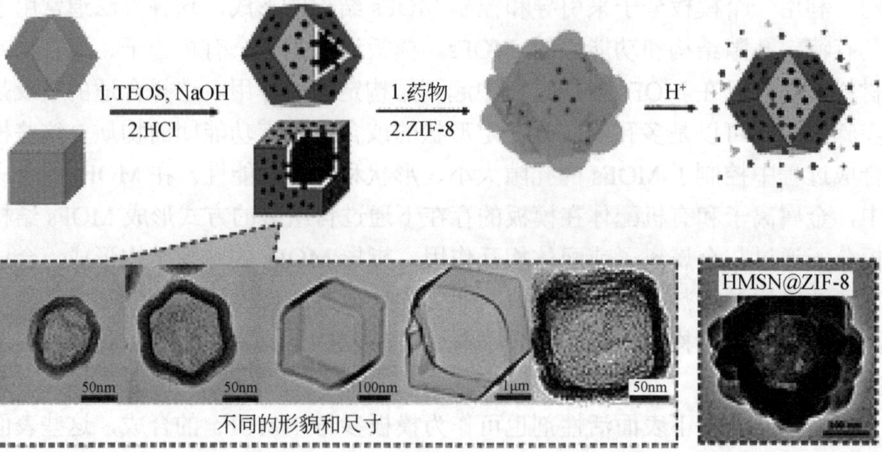

图 3.4　基于 pH 值的 HMSN 的药物递送系统[35]

Zha 等在研究 MOFs 生物催化剂的仿生矿化催化行为时，验证了脂肪酶在 MOFs 中的包埋固定化过程。降低的值和表观活化能证实脂肪酶@ZIF-8 比游离酶具有更优异的酶-底物亲和力。当脂肪酶被封装到多孔沸石状拓扑结构中时，会伴随发生构象重排，其中脂肪酶结构在特性上表现出更大的灵活性[36]。

Miao 等以一种过氧化物酶（HRP）为例，提供了一种通过仿生矿化优化生物传感器热稳定性和活性的合理设计策略。为了克服生物传感器较弱的热稳定性，Fe-MOF 的矿化在 HRP 上形成了一层保护层，可抵御高温。此外，仿生矿化 HRP@Fe-MOF 可以双重催化 TMB/H_2O_2 用于显色。由于热稳定性适体和仿生矿化 HRP@Fe-MOF，该生物传感器也可以通过简单的热处理回收[37]。

Yan 及其同事[38]展示了一种生物矿化的细菌，该细菌可将多种治疗剂输送到肿瘤部位以进行协同治疗。该报告表明，金属有机框架（MOFs）生物矿化细菌在装载治疗剂后仍能保持其生存能力和肿瘤选择性。制备的细菌@MOFs 形式具有显著的治疗功效，拓宽了基于生物体的生物矿化的应用领域。

MOFs 的生物矿化为细胞内蛋白质递送提供了一种强大的方法，使得研究蛋白质的生物功能和治疗潜力成为可能。然而，这种方法的效力在很大程度上受到当前将蛋白质与 MOFs 连接以进行生物矿化和细胞内递送的策略效率低下的挑战。有研究报告了一种多功能且方便的生物矿化策略，用于使用 MOFs 快速封装和增强蛋白质的递送[39]。研究结果表明，富含组氨酸的绿色荧光蛋白可以通过促进蛋白质和金属离子之间的协调来加速 MOFs 的生物矿化，从而使蛋白质递送效率提高达 15 倍。

近期有研究开发了一种 ZIFs 纳米晶体的原位仿生矿化策略，构建一种具有良好细胞相容性、高稳定性以及 pH 响应性的药物释放体系[40]。溶菌酶（Lys）被包裹在 ZIF-8 表面，并与金属离子紧密结合，促进骨样羟基磷灰石（HAp）的成核和生长，形成 HAp@Lys/ZIF-8 复合材料。体外实验表明，具有空心 Lys/ZIF-8 核和 HAp 壳的复合材料具有较高的载药效率（56.5%）、良好的 pH 响应药物传递、细胞相容性和在生理条件下的稳定性。Miao 等模拟了生物体分泌无机矿物质以形成外骨骼 OVA 周围铝基 MOFs 的仿生矿化（OVA@Al-MOFs）来形成一个带正电荷的外骨骼笼。YCs 可以有效携带 Al-MOF-装甲抗原靶向肠道 M 细胞，并通过黏膜上皮的主要通道传递，诱导有效而持久的免疫，该过程促进了该方法在口服疫苗接种中的应用。

此外，有研究开发了一种基于 pH 敏感的沸石咪唑酯框架（ZIF）-8 的超温和的简易方法来增加孔隙率，提供最大孔径为 20 nm，是平均孔径的 8 倍[41]。这得益于生物催化过程中产生的酸性微环境，导致葡萄糖氧化酶（GOx）被引入 ZIF-8 进行生物矿化。另外还发现生物矿化策略能够产生稳定的中孔，该中孔足够大，能够用来装载乳糖酶并使其保持良好的酶活性。

4. 前驱体策略

MOFs 的前驱体化方法主要通过自组装金属含量的“节点”和有机连接剂的“桥”来构建一维、二维或三维的网络结构，这种结构具有非常高的孔隙体积和表面积。通过选择不同的金属中心和有机连接剂，可以设计 MOFs 的框架拓扑结构和孔结构及大小。它们的化学性质可以通过连接剂的化学功能化和后续修改来改变。

MOFs 材料的出现和应用，拓展了多孔金属氧化物的合成方法。以 MOFs 为前驱体合成多孔金属氧化物具有独特优势。第一，方法简单，制备过程中无需添加表面活性剂。第二，成本不高，常用的 MOFs 前驱体（如普鲁士蓝）原料便宜且产率高，多孔金属氧化物的转化过程也操作简便，转化率高。第三，可控性强，MOFs 颗粒的粒径和形貌具有高度可控性，从而可以间接控制多孔金属氧化物的粒径和形貌，可制备多种不同形貌的多孔金属氧化物，如正方体、中空立方体、球体、中空球体、八面体和纺锤体等。第四，可大量生产，常用 MOFs 前驱体的量产性确保了相应多孔金属氧化物的可量产化。总的来说，MOFs 前驱体化方法的多样化使研究人员能够根据特定应用需求定制 MOFs 材料的性能。

以 MOFs 为前驱体制备先进功能材料还有两大优势。首先，可以通过一步法制备金属氧化物/碳纳米复合材料，方法简单。由于 MOFs 材料既是碳源又是金属源，通过控制转化条件，可以同时保留这两种成分，从而得到金属氧化物/碳纳米复合材料。其次，可以通过一步法制备多金属氧化物纳米复合材料，特别是核壳结构（core-shell）多金属氧化物纳米材料。核壳结构 MOFs 材料（MOF@MOF）因其独特的结构和性能，成为 MOFs 化学的研究热点。以 MOF@MOF 作为模板，可以一步合成核壳结构的多金属纳米复合材料。

以 MOFs 材料为前驱体合成先进功能材料的研究已经取得了一些成果，这些功能材料在科研和应用中表现出了巨大的潜能。Geng 报告了一种方便的部分原位硫化策略构建 CuS@Cu-MOF 纳米复合材料作为新型多功能纳米材料，以最大限度地提高抗癌功效[42]。CuS@Cu-MOF 纳米复合材料由长 200 nm、宽 40 nm 的梭形 Cu-MOF 组成，4 nm 大小的 CuS 纳米点均匀分布在其表面或孔结构中。由于 CuS 的等离子体效应，该 CuS@Cu-MOF 纳米复合材料在光热治疗中表现出更高的近红外光吸收和较高的热转换效率（39.6%）。Yu 等通过设计钴-MOF 前驱体 CUST-591 并在氮气条件下碳化，成功合成了一种新型复合材料[43]，呈现出 2 倍的三维网相互渗透。

3.3 药物的缓控释

体内代谢和循环系统使药物浓度维持时间较短，导致血液中药物浓度波动较大。当药物浓度超过患者的耐受剂量，会产生不良反应，而低于有效剂量又无法实现疗效。为了解决这一问题，纳米药物载体通常选用水溶性好且生物相容性佳的材料，以减少在血液循环中被网状内皮系统识别并清除的概率，从而延长药物载体在体内的循环时间，使其缓慢释放药物成分，减少药物浓度波动，实现缓释长效的目的[44, 45]。

在传统纳米载体材料的基础上，人们还开发了对外部或内部刺激响应的智能药物载体。这些智能载体可以利用特殊材料（如对酸、碱、光、磁等刺激响应的材料）将药物聚集到作用部位，并在目标位置以适当浓度释放药物分子。通过修饰或功能化，这些智能载体能够避免被人体免疫系统攻陷，增强药物递送的精准性和效率。

MOFs 在缓控释递药系统中作为药物载体，能够通过多种方式避免药物的“暴发效应”，从而实现对药物释放的精确控制，并延长药物在体内的滞留时间（图 3.5）。主要实现方式包括以下几种：一是通过主客体分子间的相互作用，调控药物的释放速率；二是对 MOFs 表面进行改性，提升其药物负载能力和释放控制能力；三是通过调控 MOFs 的结构缺陷，优化其性能，进一步提高药物释放的可控性。

3.3.1 主客体分子间相互作用

利用 MOFs 与药物之间的主客体分子间相互作用，可以实现缓控释递药。药物分子通常通过氢键、π-π 堆积、配位作用和静电作用等方式负载于 MOFs 的孔隙或表面，使 MOFs 成为药物的缓控释递送载体。例如，低细胞毒性的卟啉金属有机框架 PCN-221 对甲氨蝶呤具有高载药量。甲氨蝶呤分子通过扩散进入 PCN-221 的孔隙和孔道，并通过 π-π 作用及氢键作用与 PCN-221 结合，从而在生理环境下实现缓释行为。这种方式不仅提高了药物的稳定性，还延长了其在体内的作用时间，有助于提升治疗效果[46]。

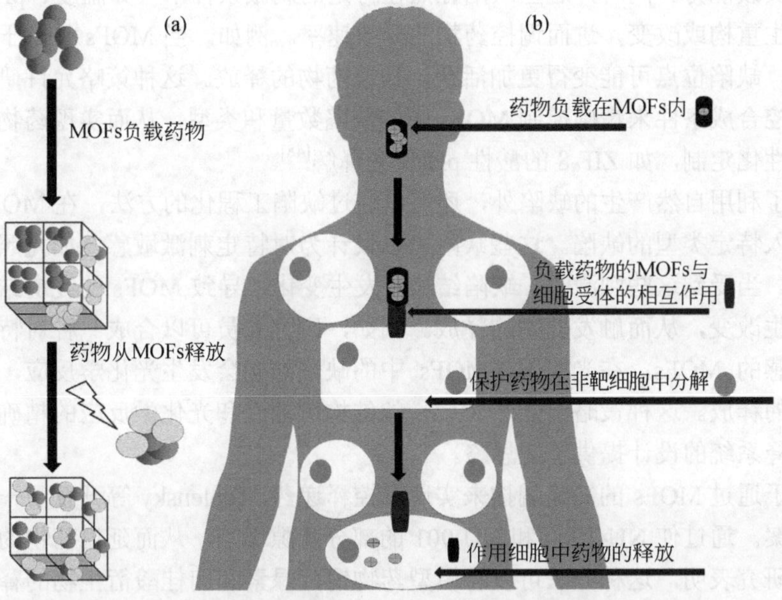

图 3.5　使用金属有机框架作为药物传递载体药物在人体内释放

3.3.2　调节孔径及对 MOFs 表面改性

利用 MOFs 的表面改性可以实现缓控释递药。人体的生理环境非常复杂，仅依赖 MOFs 与药物之间的主客体分子间相互作用，很难在复杂的生理条件下实现精确可控的药物递送。通过对 MOFs 进行表面改性，可以显著提高其作为药物缓控释载体的性能，实现精准药物递送。这种改性方法不仅能够增强 MOFs 的药物负载能力和释放控制能力，还可以改善其生物相容性和稳定性，从而确保药物在体内的高效和安全释放。

有研究表明聚乙二醇（PEG）表面涂层可以显著提高普鲁卡因胺（PA）制剂 PA@ZJU-64-NSN 在胃液中的化学稳定性。其释药过程可由内源性 Na 触发，从而在目标肠道环境下实现控制释放。对壳聚糖包衣的 5-Fu@Zr-NDC MOFs 的药代动力学分析表明，5-Fu 在酸性环境中的释放量仅为 20%，而在人工肠液中的释放量则高达 70%，显示出良好的控释效果[47]。

3.3.3　利用 MOFs 缺陷调控

在某些情况下，MOFs 的合成过程中会不可避免地产生一些缺陷，如配体缺失或金属离子空位。这些缺陷结构可以作为药物分子的吸附位点，增强 MOFs 对药物的负载能力。同时，这些缺陷位点在特定的刺激条件下（如温度、pH 变化）可能发生重构或改变，进而调控药物的释放速率。例如，当 MOFs 暴露于酸性环境中时，缺陷位点可能变得更加活跃，加速药物的释放。这种策略允许研究人员通过调控合成条件来精确控制 MOFs 中的缺陷数量和类型，从而实现药物释放行为的个性化定制，如 ZIF-8 的酸性 pH 反应降解[48]。

除了利用自然产生的缺陷外，还可以通过缺陷工程化的方法，在 MOFs 中人为地引入特定类型的缺陷。这些缺陷可以设计为对特定刺激敏感，如光照、声波或磁场。当受到这些刺激时，缺陷结构会发生变化，导致 MOFs 的孔道结构或稳定性发生改变，从而触发药物的释放。例如，研究人员可以合成一种对特定波长光照敏感的 MOFs，在光照下，MOFs 中的缺陷结构会发生光化学反应，导致药物分子的释放。这种策略结合了 MOFs 的结构可调性和光化学反应的精确性，为药物控释系统的设计提供了新思路。

基于通过 MOFs 的缺陷调控来实现缓控释递药，Teplensky 等开发了一种缺陷调控方案。通过使 NU-1000 和 NU-901 的部分孔隙塌陷，从而延迟药物的释放。他们的研究表明，这种方法可以将模型药物钙黄绿素和肉桂酸衍生物的释放时间有效延迟约 2～7 天，并在 30～49 天后实现完全释放。这种缺陷调控策略显著提高了药物的缓控释效果[49]。

3.3.4　刺激响应型 MOFs 载体及靶向给药

MOFs 具有不同的 pH 稳定性，因此，可以利用 MOFs 在靶细胞周围特殊的 pH 微环境下分解框架结构，从而释放药物，实现动态调控。例如，ZIF-90 在中性和微碱性环境中可以保持结构稳定性，当负载有阿霉素和 5-Fu 的 ZIF-90 处于肿瘤微环境（pH=5.5）下时，其药物释放量可以达到 95%以上。这种 pH 敏感性使 MOFs 成为实现精准药物递送的理想载体[50]。MIL-101（Fe）具有开放的金属位点，有研究把姜黄素吸附于这些位点使其成功包封在 MOFs 中。结果表明在酸性条件下，姜黄素的释放速率显著提高，而在正常细胞环境中则缓慢释放[51]。

有研究通过 MOFs 载体，将化疗与羟基自由基介导的化学动力学治疗（CDT）合理整合，在癌症治疗中具有巨大的潜力[52]。与 Cu-MOF 相比，pH 响应 MAF 被选择为载体，由于较高的药物负载（17.6%）和相对均匀的大小，阿霉素（Dox）作为模型药物装载。HA 壳作为一种智能靶向肿瘤的“引导剂”通过静电相互作用涂在 MAF@Dox 表面，通过与金属的配位作用获得功能化的 HA-MAF@Dox。

封装的Dox、3-AT和Cu^{2+}通过HA-MAF@Dox纳米载体在癌细胞中通过一系列复杂的化学变化缓慢释放。体外实验表明，HA-MAF@Dox具有较高的Dox转运效率，能有效调节过氧化氢酶（CAT）活性，增强了对HepG2细胞的细胞毒性，检测到HA-MAF@Dox颗粒中Dox的pH响应释放特性。结果表明，在酸性条件下Dox的释放率约为75%，而在生理条件下只有15.5%。也有研究开发一种MOFs膜伪装的多种药物传递策略，通过同时传递药物、消耗GSH和提高ROS水平，来靶向氧化应激[53]。当MOFs载药系统分散在pH为7.4的溶液中时，24 h内只有27.9%的Dox被释放。随着溶液pH值的降低，纳米颗粒的Dox释放量增加。当溶液的pH值为5.5时，纳米颗粒的Dox释放量在24 h内可达到70.3%。因此，MOFs膜伪装的多药物传递纳米平台在酸性环境下可以释放药物，而在中性环境下却很难释放药物，这一结果有望实现精确、有效的肿瘤治疗。此外也有研究直接纳米沉淀技术将天然物理系统（PHY）载药封装到沸石咪唑酯框架（ZIFs）中合成PHY@ZIF-8。在酸性（pH=5.0）培养基中，PHY@ZIF-8的药物释放行为比pH为7.4的生理培养基高3倍[54]。

Schnabel等[55]报道了一种称为Zn-MOF-74的金属有机框架，它包含Zn(Ⅱ)离子和2,5-二羟基苯-1,4-二羧酸酯配体，作为三氧化二砷的药物纳米载体。由于Zn-MOF-74中存在开放的金属位点，因此可以向该材料中装载大量的As(Ⅲ)-药物(以亚砷酸的形式)。此外，他们还研究了在不同pH值下的药物释放，As(Ⅲ)的释放是通过pH触发的，在pH=6.0时比在pH=7.4时更快。Akbar等[56]开发了用叶酸缀合壳聚糖（FC）修饰的多功能双金属MOFs作为药物递送系统（DDS），用于靶向递送5-氟尿嘧啶。实验结果观察到pH响应性药物释放，58%的负载5-氟尿嘧啶在模拟pH（5.2）的癌细胞中释放，而在生理pH（7.4）下其仅释放24.9%。

3.4 装载策略

3.4.1 直接装配策略

原位包装法适用于小于MOFs孔的药物，以及人体内可能过早释放的制剂，也称为“一锅法”或直接装配法。在“一锅法”中，药物被视为MOFs的组成部分，并参与了纳米平台的合成步骤。原位包装的优点是缩短了封装时间，对药物的大小没有限制，可以容纳大尺寸的药物。根据负载药物的结构、性质和官能团，药物与MOFs之间的相互作用丰富，如静电作用、氢键、π-π作用、化学吸附、范德瓦耳斯力作用等。上面提到的那些交互作用可以减少药物在MOFs基质中的流动性，能够实现其在体内缓释的目的。

在用于药物递送的MOFs中，直接装配可以将药物封装在MOFs的孔隙中。

孔隙封装策略涉及将治疗剂掺入 MOFs 的多孔结构中。该技术利用 MOFs 的高表面积和可调孔径来实现受控的药物负载和释放。通过物理吸附或化学相互作用，药物被限制在孔隙内，从而在特定的生理条件下实现持续释放。这种方法优化了治疗效果，同时最大限度地减少了副作用，使 MOFs 成为先进药物递送系统的有前途的平台。

一般来说，表面吸附是通过在功能分子溶液中搅拌预先合成的 MOFs 来实现的。由于高表面积和孔隙率，功能分子可以吸附在 MOFs 表面。范德瓦耳斯力、π-π 键相互作用和分子间氢键是该方法的主要驱动力。这种相对简单的策略对 MOFs 的孔径或官能团类型没有严格要求，然而，由于分子与 MOFs 框架之间的相互作用较弱，浸出问题难以避免。为防止 MOFs 纳米颗粒在体外和体内生理条件下的解离，Liu 团队开发了一种针对 MOFs 纳米颗粒的直接封装策略，以增强生理条件下的稳定性和刺激响应的细胞内药物释放。将包裹在表面的聚合物作为屏蔽，纳米级 MOFs 不被磷酸根离子或酸分解，防止装载的药物泄漏[57]。

采取“一锅法”将金属离子、有机配体和目标分子同时加入溶剂中，可以使 MOFs 晶体在形成过程中封装目标分子。尽管这种方法能够实现目标分子的高效包埋，并且对分子尺寸没有严格要求，对蛋白质和核酸等大分子也有良好的包埋效果，但需要在温和条件下进行，因此适用的 MOFs 材料有限[58]。有研究将三种酶包埋在 ZIF-8 中，实现了三酶的串联反应。Alsaiari 等也使用 ZIF-8 来封装基因组编辑器 CRISPR/Cas9（蛋白质/RNA），提高其基因编辑效果，其中质子化的咪唑有利于 CRISPR/Cas9 在细胞内释放[59]。

Wang 等以铈盐和 2-甲基咪唑为原料，通过“一锅法”实现了高剂量的二氢卟吩 e6（chlorin e6，Ce6）和阿霉素（doxorubicin hydrochloride，Dox）的非均相负载得到 MOF/C&D，克服了 Ce6 在高浓度时的自猝灭现象，实现了非均相负载的高浓度的 Ce6 的光热和光动力学治疗[60]。Eddaoudi 等在 2008 年首次成功通过原位形成的卟啉修饰 MOFs[61]。采用“一锅法”将卟啉包裹在 MOFs 内部，小孔隙可以防止卟啉的泄漏，并促进小分子的扩散。

Wang 等研究采用一锅法合成 Dox@Cu/ZIF-8，并作为模板。然后，在模板中通过配体交换，获得 MAF@Dox[11]。然后，将透明质酸作为靶向配体通过静电相互作用涂覆在 MAF@Dox 表面，以降低抗癌药物对正常细胞的毒性，具有 pH 响应性，该平台通过有效调节过氧化氢分解为水和氧的关键酶过氧化氢酶的活性来增强化学动力学治疗（CDT）的疗效。还有研究人员将晚期糖基化终末产物（RAGE）抑制剂受体装入基于钴（Co）的 MOFs（ZIF-67）中，通过一锅法合成载药纳米颗粒（图 3.6）[62]。

Rezaee 等采用“一锅法”将 Zr-MOF 合成到羊毛织物中。将丹参和丹参提取物装载到改性织物上，其吸收能力分别提高了 1154%和 1842%。合成的 MOFs 对大肠杆菌的抑菌活性均为 100%，对金黄色葡萄球菌的抑菌活性分别为 60.95%和 64.64%[63]。Duan 等把介孔二氧化硅纳米颗粒用于模型疫苗卵清蛋白的装载[64]，也通过一步法合成了具有 pH 响应能力的 MOFs。

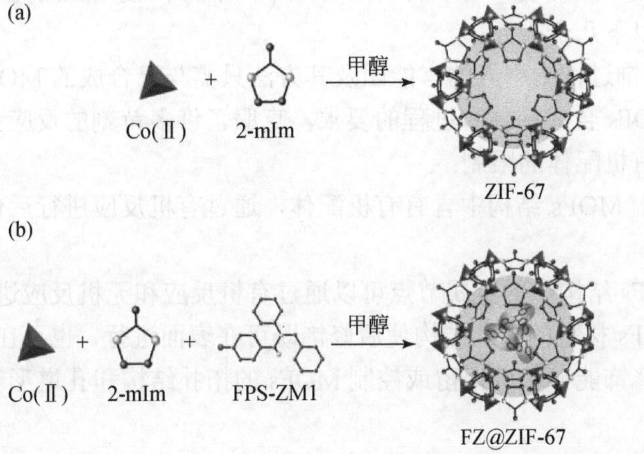

图 3.6 （a）ZIF-67（a）和 FZ@ZIF-67（b）NPs 的合成示意图[62]

Chen 等以“一锅法”成功制备了高荧光的空心[4-(4-羧基苯基)苯基]乙烯（TCBPE）MOF 纳米管，这种空心六角形纳米管可以以 pH 依赖的方式释放药物，以响应特定的固化环境，通过药物的逐渐释放，在一段时间内进一步杀死肿瘤细胞[65]。Cong 等设计了一种具有良好荧光特性和靶向给药癌症治疗系统。由于 Nd 的优异 NIR 光学特性，该系统具有生物成像的潜力[66]。

Liu 等也报道了一种原位直接封装策略，将阿霉素（Dox）封装的沸石咪唑酯框架（ZIF-8）组装在 Zr(Ⅳ)基卟啉 MOFs 表面（ZIF-8）。原位封装的方法在 ZIF-8 层的厚度上提供了较高的可控性，且不需要对卟啉类 MOFs 进行任何后续的表面修饰[67]。Peng 等设计并制造了一个金属有机框架-红细胞（RBC）膜伪装多药递送纳米平台，用于联合铁-凋亡治疗多药耐药癌症。通过在一锅溶剂热合成过程中与金属簇配位，将铁凋亡和化疗药物嵌入铁（Ⅲ）基 MOFs 中心部位的缺陷位点。红细胞膜可以伪装纳米平台，以实现更长时间的循环[53]。

Dong 等采用一锅法合成 Dox@Cu/ZIF-8，并作为模板，在模板中通过配体交换，获得 MAF@Dox，将透明质酸作为靶向配体（HA）通过静电相互作用涂覆在 MAF@Dox 表面，以降低抗癌药物对正常细胞的毒性[52]。

直接装配法是一种简单快捷的药物装载方法，实验操作流程简单，目前已实现多种药物的封装，但是仍存在诸多问题有待解决。例如，“一锅法”可以成功装载药物但并不能保证药物负载的稳定性，其次，目前 MOFs 负载多种药物的原位装配法研究较少，药物装载的有效性有待考究。

3.4.2 后修饰策略

后修饰是近年来发展起来的一种对已合成的超分子配合物进行进一步化学修饰的过程（图 3.7）。在 MOF 材料的合成中，后修饰技术应用尤为广泛，其优势主要体现在以下几个方面：

（1）后修饰过程中引入的官能团及其方法只需与已合成的 MOFs 产物相匹配，无需考虑 MOFs 合成前体和过程的要求，克服了许多苛刻的反应条件（如水热条件）以及对有机配体的限制；

（2）由于 MOFs 结构中含有有机配体，通过有机反应进行后修饰的选择性非常广泛；

（3）MOFs 结构中的金属节点可以通过有机反应和无机反应进行修饰和改变；

（4）MOFs 材料的多孔结构使后修饰既可在表面进行，也可在孔洞内部进行；

（5）后修饰能够精确保留或控制 MOFs 的拓扑结构和孔道形貌。

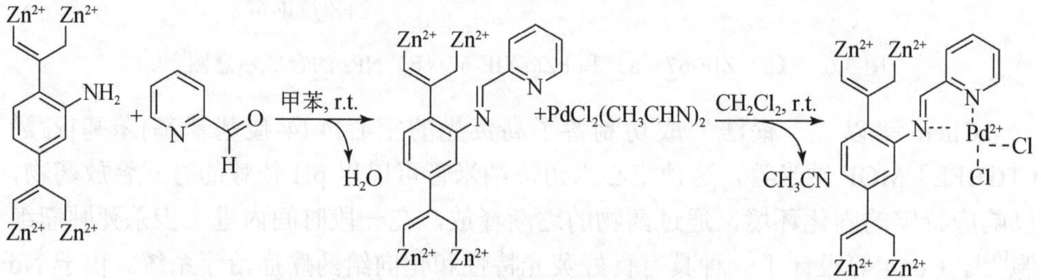

图 3.7　共价和配位后修饰结合的例子[68]

早期的 MOFs 后修饰工作主要基于 HKUST-1 和 MOF-5 这两种经典材料。随着 MOFs 应用开发的不断推进，具有更高热稳定性和化学稳定性的 MOFs 材料，如 UiO-66、MIL-88、MIL-101 和 MIL-125，已成为后修饰研究的主要材料。这种方法可分为两个过程。首先，需要将 MOFs 纳米颗粒完全合成，以获得一个骨架结构。其次，药物通过共价或非共价结合反应被封装到预先合成的骨架结构中。由于 MOFs 固有的尺寸和孔隙结构，比框架尺寸小的药物更容易包装。由于使用方便，后包封法已被广泛用于将各种药物装载入多孔 MOFs 中。MOFs 的合成条件（如高温、有机溶剂和酸性环境）对酶等生物分子过于苛刻，导致其无法维持结构特征和活性。为了解决这个问题，研究者们采用合成后修饰策略，对 MOFs 进行孔隙包裹。这种方法提供了一条在温和条件下合并生物分子的有效途径，确保了生物分子的结构和功能得以保持。

虽然后封装方法简单经济，但反应过程较长。此外，包封过程需要过量的药物，并要求药物的尺寸足够小，以保证其能够顺利通过 MOFs 的孔。例如，Zhao

和他的同事开发了一种稳定的方法，基于柔性化学修饰的 ZIF-8 涂层（CPT-CuS-ZIF-8@HA）。在晶体生长过程中，将硫化铜纳米颗粒嵌入 ZIF-8 框架中，将喜树碱（CPT）原位捕获在混合框架的孔隙中，一步得到 CPT-CuS -ZIF-8。CPT-CuS-ZIF-8@HA 是通过配位键和静电相互作用修饰 CPT-CuS-ZIF-8 的透明质酸（HA）表面。在后续研究中发现，局部给药、智能和时空药物释放以及化疗和光热治疗（PTT）的协同组合都是该技术的明显优势。

Alves 研究组用介孔 N_3-Bio-MOF-100 装载一种天然化合物姜黄素作为抗癌药物，表面通过点击化学与 FA 基团功能化，使用合成后修饰来选择性地靶向和杀死乳腺癌细胞（图 3.8）[21]。有报道用合成的环糊精金属有机框架（CD-MOFs）装载异硫氰酸烯丙酯（AITC）作为干粉吸入，以增强肺部药物输送[69]。

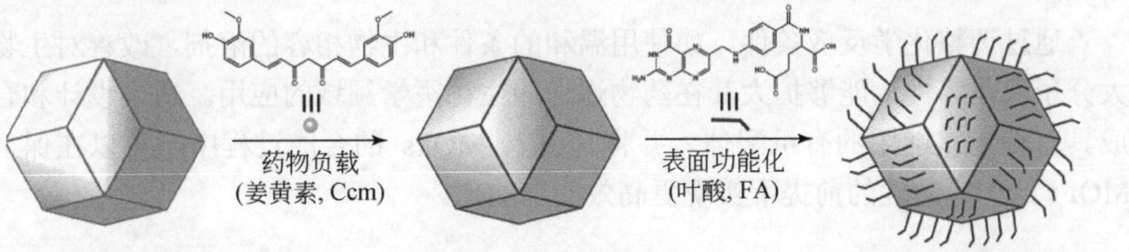

N_3-Bio-MOF-100　　Ccm@N_3-Bio-MOF-100　　Ccm@N_3-Bio-MOF-100/FA

图 3.8　利用合成后修饰开发肿瘤靶向载药 MOFs 的合成途径[21]

Ma 等首先用锆（Ⅳ）氯离子和苯-1,4-二羧酸合成了 Zr 基 MOF UiO-66。然后，用合成后装载的方法封装姜黄素。光谱和热重分析表明，该系统装载率达到了 3.45%（质量分数）[70]。Nam 等用合成的 Cu-MOF 固载 PDMS（聚二甲基硅氧烷），合成的材料对五种菌株（大肠杆菌、金黄色葡萄球菌、铜绿假单胞菌、肺炎克雷伯菌和耐甲氧西林金黄色葡萄球菌）表现出浓度依赖的抗菌活性。此外，PDMS@Cu-MOF 保持了 PDMS 的物理和热特性，并对小鼠胚胎成纤维细胞具有较低的细胞毒性[71]。Shan 等利用 1,3,5-苯三甲酸（Fe-MIL-100）作为杀菌剂偶氮基菌素的载体。由于其比表面积高达 2251 m^2/g，Fe-MIL-100 的装载含量令人满意，可达 16.2%[72]。Alijani 等利用四氧化三铁核心和 UiO-66-NH_2 壳层的纳米结构（Fe_3O_4@MOF）来装载抗癌药物阿霉素。他们将负载的纳米结构偶联到高荧光碳点（CDs）上，并结合配体 AS1411（适配体）进行覆盖。该 Fe_3O_4@MOF-DOX-CDs-Apt 纳米载体在过表达癌细胞的环境下被特异性解锁，完成药物的释放[73]。

通过后修饰，可以在 MOFs 结构中引入各种官能团，如催化剂、药物、荧光标记等，使 MOFs 能够针对特定的应用进行优化，提供了一种灵活的方法来调整 MOFs 的化学性质，使其无需在 MOFs 合成的初期阶段引入复杂的官能团，简化了合成过程。后修饰过程通常不破坏 MOFs 的晶体结构，保持了其孔隙结构和表面积，这对于保持其储存和分离性能至关重要。然而后修饰通常涉及在已形成的

MOFs 孔隙中引入分子，这可能限制大分子的装载或导致荷载效率不高。某些反应需要特定的反应条件，如特定的溶剂或温度，这可能限制其在靶向触发中的使用。基于弱相互作用力（如氢键、范德瓦耳斯力等）的后修饰可能在生理条件下不稳定，容易导致封装物质的早期释放。

这就需要开发新的化学连接策略，如共价键合，提高荷载分子的稳定性和 MOF 的功能性。例如，通过使用点击化学反应在 MOFs 表面或内部结构中引入功能分子。Renata 等讨论了一种策略，利用“点击”化学将功能性叶酸（FA）分子锚定在 N_3-Bio-MOF-100 的表面上，从而实现靶向肿瘤的载药 MOFs。使用姜黄素作为抗癌药物。因此，“点击”化学已被证明在药物输送中非常有效。但由于大量具有各种特征的“点击”反应的可用性，为给定应用选择合适的修饰还需进一步研究。

通过调整化学反应条件，如使用温和的条件和生物相容的溶剂，改善对生物大分子的兼容性，能够扩大其在药物递送和生物医学领域的应用。适当设计和合成具有特定官能团的有机配体，并将其引入 MOFs 的合成过程中，可以在保证 MOFs 结构完整性的前提下实现更高效的功能化。

3.5 毒理学研究

3.5.1 MOFs 材料的生物安全性

纳米药物临床应用的一个重要问题是生物安全。因此，基于 MOFs 系统的 NPs 毒性需要进行更系统的体内研究。确保生物安全的第一步是通过平衡无机和有机设备来建立 MOFs。无毒或低毒性的物质需要在体内和通过体内的代谢系统产生安全的分解产物。此外，基于 MOFs 的控释药在给药后采用 ADME（吸收、分布、代谢和排泄）机制，在生物体内进行的体内研究中常用的小动物模型并不能预测药物的疗效和药物控释系统在人体中的性能。因此，必须建立创新的体内模型来准确地预测人体内部药代动力学的特征。这将意味着这种控释的成功是临床转化的巨大一步。

基于 MOFs 的 DDS 临床应用的另一个主要挑战是其潜在的毒性。然而，现有的文献非常有限，无法得出关于 MOFs 纳米颗粒毒性的结论。到目前为止，已经对不同的细胞系进行了许多体外毒性研究，这使得比较所获得的结果非常模糊。根据 Baati 等的研究，MOFs 纳米颗粒可以在尿液或粪便中进行进一步的生物降解和消除，而不发生代谢，引起显著毒性[74]。MOFs 的生物相容性仍然是限制其发展的关键因素之一。在水溶液或生理条件下，MOFs 的稳定性较差，容易迅速聚集或快速崩解，导致细胞凋亡和组织异常等副作用。因此，如何在引入多种治疗方式的同时提高 MOFs 的生物相容性，已成为 NMOFs 领域研究的重点问题。

3.5.2　金属离子及配合物毒性

MOFs 的毒性是临床研究前必须解决的首要问题。由于 MOFs 的结构和种类多样，以及生物体内部环境的复杂性，MOFs 的毒性不仅与其结构、组成、形态、大小和稳定性有关，还与活体组织的耐受性有关。因此，需要对各种 MOFs 的毒性进行综合评估。使用 MOFs 进行生物医学应用需要其具有良好的生物友好性，但目前关于 MOFs 或配位聚合物的毒性结果数据大多局限于金属和有机配合物的毒性评价（表 3.1）。已报道的毒性研究主要集中在体外或/和体内急毒实验，而体内长期毒性，特别是 MOFs 框架体内系统的吸收、分布、代谢和排泄过程研究很少。为了全面确定 MOFs 的毒性，迫切需要广泛的体内研究和长期的组织积累监测。

表 3.1　口服 LD_{50}（大鼠）和每日最大摄入量（人）

金属元素	LD_{50}/（g/kg）	每日最大摄入量/mg
Zr	4.1	0.05
Ti	25	0.8
Cu	0.025	2
Mn	1.5	5
Fe	0.45	15
Zn	0.35	15
Mg	8.1	350
Ca	1	1000

MOFs 在 DDS 中使用时，应首先进行生物相容性和毒性测试。利用内源性或生物活性分子作为配体，结合高生物相容性的金属离子（如铁、钙、锌等）作为金属节点来构建功能性 MOFs，有助于避免 MOFs 的毒性。MOFs 是由金属团簇和有机连接物构成的，因此应考虑金属和连接物的毒性。在研究过程中可通过考虑中位致死剂量（LD_{50}）来评价金属的毒性。给药最推荐的金属是镁、钙、锌、锆、铁和钛，其口服 LD_{50}（g/kg）分别为 8.1、1、0.35、4.1、0.45 和 25。另一方面，许多生物活性物质如酸、肽、蛋白质前药等已被用作合成 MOFs 的有机配合连接体[75]。内源性配体比外源性配体更具有生物相容性，如天冬氨酸基配体、腺嘌呤和环糊精。但从目前已有报道来看，最常见的连接物是外源性连接物，如甲酰羧酸盐和咪唑酸盐。通过研究发现，大鼠口服对苯二甲酸、三聚酸、2,6-萘二羧酸和 1-甲基咪唑的口服剂量分别为 1.13 g/kg、5.5 g/kg 和 8.4 g/kg。

在研究 MOFs 配体毒性时，亲水-疏水平衡也是一个重要参数。研究表明，含有亲水基团的 MOFs 具有相对较低的细胞毒性。比如，Zr-MOF（UiO-66）因其优异的化学和热稳定性以及重要的孔隙率而受到人们的极大关注。且锆是一种低毒性金属（口服致死剂量 LD_{50} 约为 4.1 g/kg）。在目前的研究中，斑马鱼被广泛用作研究环境毒性的优良模型。Ruyra 等报道了一些经典的未改性 MOFs 纳米粒子在体外对 HepG2 和 MCF7 细胞的影响，并研究了它们在斑马鱼胚胎体内的毒性[76]。纳米 MOFs 的毒性试验研究结果表明，Mg-MOF74、Co-MOF74 和 MIL100 在斑马鱼胚胎体内的毒性最低。作为药物传递的一种有吸引力的载体，在使用前考虑 MOFs 的毒性是很重要的。因此 Wuttke 等[77]对不同类型的纳米颗粒（MIL-100-(Fe)、MIL-101(Cr)和 Zr-Fum）的生物用途安全性进行了深入研究。评价了这些 NMOFs 对小鼠肺和人内皮细胞的影响及其毒性。

此外，钛基 MOFs 由于其具有较好的生物相容性和溶解度而在抗菌治疗、癌症治疗、炎症治疗、口服治疗和骨损伤治疗等临床研究中得到了广泛的应用。从抗菌方面来看，装载抗菌药物的钛基 MOFs 可以通过光催化抑制产生活性氧的细菌生物膜的形成，以辅助提高抗菌活性。在癌症治疗中，已经证明了钛基 MOFs 的改性表面具有更高的稳定性和更高的孔隙率，这可以进一步提高载药能力。在对抗炎症方面，钛基 MOFs 有效地解决了抗炎药物的携带和控制释放问题。在口服给药方面，钛基 MOFs 作为药物载体是安全的，药物释放后可完全排泄。

3.5.3　生物相容性和生物降解性

体内纳米载体药物的活性药物成分及载体材料主要通过肝脏和其他组织中的代谢酶进行代谢。此外，载药粒子易被单核吞噬系统（MPS）吞噬，进而被溶酶体降解或代谢，这可能影响药物和载体材料代谢及降解产物的种类和数量。因此，研究药物代谢需要确定活性药物和载体材料的主要代谢和降解途径，并对其代谢及降解产物进行分析（表 3.2）。

表 3.2　已报道 MOFs 生物相容性研究

MOFs	测试细胞	结果	参考文献
Zr-MOF@NAP/AV	人成纤维细胞（HFFF2）	具有良好的细胞相容性	[5]
DOX@Fe_3O_4/Bio-MOF-13 核/壳复合材料		DOX@Fe_3O_4/Bio-MOF-13 核/壳复合材料似乎是一种安全、经济、高效的方式治疗乳腺癌的可行选择	[78]
DUCNP@Mn-MOF/FOE	小鼠乳腺癌模型	DUCNP@Mn-MOF/FOE 可抑制肿瘤生长，但无显著毒性	[79]
γ-CD-MOF		具有良好的肺局部耐受性、通透性，无明显毒性	[69]

续表

MOFs	测试细胞	结果	参考文献
Cu-MOF（PDMS@Cu-MOF）	小鼠胚胎成纤维细胞	较低的细胞毒性，正常浓度下PDMS@Cu-MOF处理的细胞活力（>95%）	[71]
CpG/ZANPs		显著抑制了EG7-OVA肿瘤的生长，同时显示出最小的细胞毒性	[80]
Fe_3O_4@MOF-DOX-CDs-Apt	HUVEC细胞	对MB-231癌细胞诱导选择性凋亡（24h>为77%）；而正常HUVEC细胞诱导凋亡不到10%	[73]
（Dox/Cel/MOFs@Gel）	（如A549，HepG2，KB，L929，L-02）成年大鼠	这种局部治疗显著降低了全身毒性，且对其他器官没有明显的损伤。MOFs的生物相容性试验表明了合理的生物安全性，没有明显的永久性毒性作用的迹象（图3.9）	[81]
hydroMOFs	VERO细胞系	细胞毒性筛选表明，hydroMOFs是一种生物相容性化合物	[82]
MTX-TA/Fe^{3+}@HA	CIA鼠模型	MOFs可通过增强甲氨蝶呤的抗风湿活性，同时通过靶向给药来降低其毒性作用	[12]
LND-HA@ZIF-8@Lf-TC		与生物体没有任何显著的相互作用	[83]
PEG–FA-NH_2-Fe-BDC	MCF-7细胞	MTT的研究表明，这些纳米载体在较低的浓度下是无毒的	[84]
UiO-66-NH_2-load and UiO-66-NO_2-load 酮洛芬Ketoprofen	软骨组织	合成的MOFs载体具有一定的生物安全性，可以作为药物传递载体	[85]

对于耐降解和代谢的无机非化学材料，通常认为它们在体内有较长的停留时间。例如，聚乙二醇化量子点可以在体内保留至少两年。然而，一些研究人员不同意这一观点，他们认为溶酶体的酸性环境可以导致无机纳米材料降解并释放金属离子，这些金属离子随后与不同的生物分子结合。此外，无机纳米材料表面的化学基团可以通过酶或非酶途径进行代谢。相比之下，有机纳米材料的酶降解时间和过程取决于其化学组成和物理化学性质。有机纳米材料可能先分解，然后代谢为更小的粒子。在肝脏中，如果材料尺寸过大，无法通过跨细胞肝窦内皮细胞之间的孔隙，而小粒径材料则可以穿过孔隙进入窦周间隙，然后进入肝细胞，随后通过单加氧酶、转移酶、酯酶和环氧化物水解酶等代谢。产生的代谢物可以通过尿液排出，或被输送到胆汁中，最终通过粪便排出体外。总的来说，纳米粒子可能产生大小、形状和化学形式不同的各种代谢物。研究这些代谢物及其路径对于理解纳米药物载体的体内行为至关重要。

排泄是减少纳米载体在体内潜在危险的基本过程。纳米药物载体中的活性药物和载体材料可能通过肾小球过滤和肾小管分泌进入尿液排泄，或通过肝脏以胆汁分泌形式随粪便排泄。载药粒子自身一般不易通过上述途径直接排泄，需要解聚成载体材料或载体材料降解后经肾脏排泄。肾脏清除通过肾小球进行过滤，该过程受纳米粒形状、大小和电荷的影响。一般来说，直径小于 6 nm 的颗粒可以通过肾小球，进入膀胱并在尿液中排出。表面电荷会影响直径在 6～8 nm 的纳米粒子的排泄，因为正电荷粒子更容易通过肾小球。透过肾小球毛细血管进入肾小囊腔，一些微粒会被再吸收并重新转化为血液，未被再吸收的颗粒则通过尿液排出。

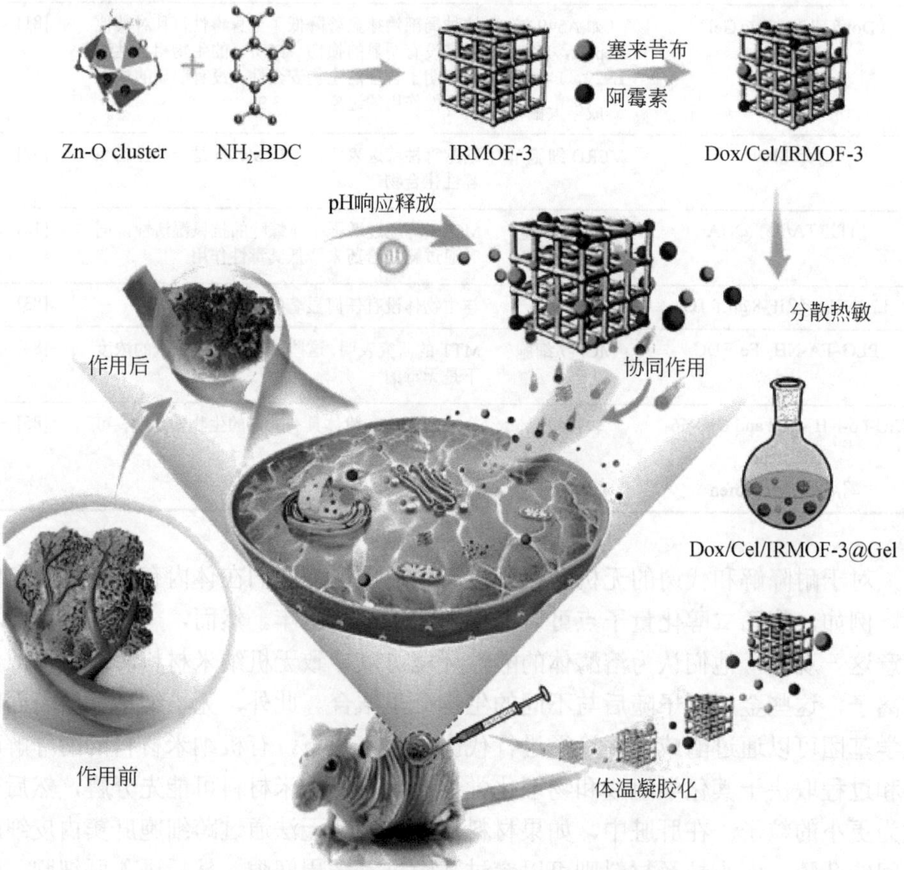

图 3.9 口服 MOFs 载药系统抗癌代谢研究[81]

据报道，对于水溶性胶束聚合物，肾清除是其主要清除途径之一。分子质量低于 5000 Da 的化合物可以自由通过肾小球进入尿液，分子量的差异会导致清除速率的不同，同一种聚合物分子量越低，越容易通过肾小球过滤排泄。直径大于 6 nm 的非生物降解纳米载药颗粒的排泄物会进入胆汁，低分子量的材料通过活性转运被处理，而高分子量的则通过顺细胞转运和跨细胞运输。不同分子量组分还可以同时经过肝脏，生成的胆汁排泄进入肠道，最终通过粪便排出体外。

药物运输递送途径、体内分布及代谢问题、纳米载体潜在的安全性问题，均与纳米粒载体及其降解产物在体内的药代动力学行为密切相关。因此，纳米载体的药代动力学研究是十分必要的。目前合成的用于药物递送领域的 MOFs 一般都拥有较好的生物相容性。

有研究检测了 PCN-224 在人肝细胞 L-02 细胞和小鼠巨噬细胞 RAW264.7 中的一般细胞毒性以及 L-02 细胞中炎症和自噬的影响。结果表明，PCN-224 被 L-02 细胞吞噬，进而导致 L-02 细胞发生形态学改变、细胞膜破坏和氧化应激。PCN-224 可能通过促进肿瘤坏死因子（TNF-α）和白细胞介素（IL-6）等炎症因子的分泌来触发炎症反应，并诱导自噬小体积累，进而导致自噬功能障碍。

参 考 文 献

[1] Liu Y, Li M, Yang F, et al. Magnetic drug delivery systems. Science China-Materials, 2017, 60: 471-486.

[2] Xu R H, Xie Q W, Li X J, et al. Modified quechers method based on multi-walled carbon nanotubes coupled with gas chromatography-tandem mass spectrometry for the detection of 10 pyrethroid pesticide residues in tea. Chinese Journal of Chromatography, 2022, 40: 469-476.

[3] Li Jingpeng, Lu Yun, Wang Huiqing. Eco polymeric materials and natural polymer. Polymers, 2023, 15(19): 4021.

[4] Hu Yuling, Lu Haibin, Yuan Xihang, et al. The histologic reaction and permanence of hyaluronic acid gel, calcium hydroxylapatite microspheres, and extracellular matrix bio gel. Journal of Cosmetic Dermatology, 2023, 22: 2685-2691.

[5] Nabipour H, Rohani S. Zirconium metal organic framework/aloe vera carrier loaded with naproxen as a versatile platform for drug delivery. Chemical Papers, 2023, 77: 3461-3470.

[6] Khatamian M, Yavari A, Akbarzadeh A, et al. Synthesis and characterization of mfi-type borosilicate zeolites and evaluation of their efficiency as drug delivery systems. Materials Science and Engineering: C, 2017, 78: 1212-1221.

[7] Nakagita T, Taketani C, Narukawa M, et al. Ibuprofen, a nonsteroidal anti-inflammatory drug, is a potent inhibitor of the human sweet taste receptor. Chemical Senses, 2020, 45: 667-673.

[8] Bazzazan S, Bidgoli K, Lalami Z A, et al. Engineered UiO-66 metal-organic framework for delivery of curcumin against breast cancer cells: An *in vitro* evaluation. Journal of Drug Delivery Science and Technology, 2023, 79: 104009.

[9] Gao Yanfeng, Wang Yanping, He Bangshun, et al. An enzyme-loaded metal-organic framework-

assisted microfluidic platform enables single-cell metabolite analysis. Angewandte Chemie International Edition, 2023, 62: e202302000.

[10] Bai Y, Dou Y B, Xie L H, et al. Zr-based metal-organic frameworks: Design, synthesis, structure, and applications. Chemical Society Reviews, 2016, 45: 2327-2367.

[11] Wang Huanhuan, Li Tong, Li Jiawei, et al. One-pot synthesis of poly(ethylene glycol) modified zeolitic imidazolate framework-8 nanoparticles: Size control, surface modification and drug encapsulation. Colloids and Surfaces A: Physicochemical and Engineering Aspects, 2019, 568: 224-230.

[12] Guo Lina, Chen Yang, Wang Ting, et al. Rational design of metal-organic frameworks to deliver methotrexate for targeted rheumatoid arthritis therapy. Journal of Controlled Release, 2021, 330: 119-131.

[13] Cao Jian, Li Xuejiao, Wang Xinxin, et al. Surface pegylation of MIL-101(Fe) nanoparticles for co-delivery of radioprotective agents. Chemical Engineering Journal, 2020, 384: 123363.

[14] Olivan R, Cocero M J, Coronas J, et al. Supercritical CO_2 encapsulation of bioactive molecules in carboxylate based MOFs. Journal of CO_2 Utilization, 2019, 30: 38-47.

[15] Zeng Jin-Yue, Zhang Ming-Kang, Peng Meng-Yun, et al. Porphyrinic metal-organic frameworks coated gold nanorods as a versatile nanoplatform for combined photodynamic/photothermal/chemotherapy of tumor. Advanced Functional Materials, 2018, 28: 1705451.

[16] Afshar E A, Taher M A, Maleh H K, et al. Magnetic nanoparticles based on cerium MOF supported on the MWCNT as a fluorescence quenching sensor for determination of 6-mercaptopurine. Environmental Pollution, 2022, 305: 119230.

[17] Ali E, Saedi F, Hussein S A, et al. Fabrication and characterization of novel nanocomposite containing Zn-MOF/gentamicin/oxidized chitosan as a highly effective antimicrobial agent. Journal of Inorganic and Organometallic Polymers and Materials, 2024, 6: 1-11.

[18] Horcajada P, Chalati T, Serre C, et al. Porous metal-organic-framework nanoscale carriers as a potential platform for drug delivery and imaging. Nature Materials, 2009, 9: 172.

[19] Akbari M, Ghasemzadeh M A, Fadaeian M. Synthesis and application of ZIF-8 MOF incorporated in a TiO_2@chitosan nanocomposite as a strong nanocarrier for the drug delivery of acyclovir. Chemistryselect, 2020, 5: 14564-14571.

[20] Pisklak T J, Macías M, Coutinho D H, et al. Hybrid materials for immobilization of Mp-11 catalyst. Topics in Catalysis, 2006, 38: 269-278.

[21] Alves R C, Schulte Z M, LuizM T, et al. Breast cancer targeting of a drug delivery system through postsynthetic modification of curcumin@N_3-Bio-MOF-100 *via* click chemistry. Inorganic Chemistry, 2021, 60: 11739-11744.

[22] Chen Lei, Yu Huimin, Li Yu, et al. Fabrication of a microporous Dy(III)-organic framework

with polar channels for 5-Fu (fluorouracil) delivery and inhibiting human brain tumor cells. Structural Chemistry, 2018, 29: 1885-1891.

[23] Wang Qianqian, Zhang Shuo, Deng Zhuang, et al. Preparation of curcumol-loaded magnetic metal-organic framework using supercritical solution impregnation process. Microporous and Mesoporous Materials, 2023, 357: 112612.

[24] Cao Shilin, Yue Dongmei, Li Xuehui, et al. Novel nano-/micro-biocatalyst: soybean epoxide hydrolase immobilized on UiO-66-NH_2 MOF for efficient biosynthesis of enantiopure (R)-1, 2-octanediol in deep eutectic solvents. ACS Sustainable Chemistry & Engineering, 2016, 4: 3586-3595.

[25] Zimpel A, Preiß T, Röder R, et al. Imparting functionality to MOF nanoparticles by external surface selective covalent attachment of polymers. Chemistry of Materials, 2016, 28: 3318-3326.

[26] Miras H N, Nadal L V, Cronin L. Polyoxometalate based open-frameworks (POM-OFs). Chemical Society Reviews, 2014, 43: 5679-5699.

[27] He J. Metal-organic frameworks as drug delivery systems for tumor targeted treatment. Highlights in Science, Engineering and Technology, 2023, 73: 62-72.

[28] Moharramnejad M, Ehsani A, Shahi M, et al. MOF as nanoscale drug delivery devices: Synthesis and recent progress in biomedical applications. Journal of Drug Delivery Science and Technology, 2023, 81: 104285.

[29] Ettlinger R, Sönksen M, Graf M, et al. Metal-organic framework nanoparticles for arsenic trioxide drug delivery. Journal of Materials Chemistry B, 2018, 6: 6481-6489.

[30] Walker G C, Konda S S M, Maji T K, et al. Preface to the "Metal-organic frameworks: Fundamental study and applications" joint virtual issue. Langmuir, 2020, 36: 14901-14903.

[31] Zlotea C, Campesi R, Cuevas F, et al. Pd nanoparticles embedded into a metal-organic framework: Synthesis, structural characteristics, and hydrogen sorption properties. Journal of the American Chemical Society, 2010, 132: 2991-2997.

[32] Zhang Xiaolei, Li Sumei, Chen Sha, et al. Ammoniated MOF-74(Zn) derivatives as luminescent sensor for highly selective detection of tetrabromobisphenol A. Ecotoxicology and Environmental Safety, 2020, 187: 109821.

[33] Liu Jiang, Bao Tianyi, Yang Xiya, et al. Controllable porosity conversion of metal-organic frameworks composed of natural ingredients for drug delivery. Chemical Communications, 2017, 53: 7804-7807.

[34] Tang Lei, Shi Jiafu, Wang Xiaoli, et al. Coordination polymer nanocapsules prepared using metal-organic framework templates for pH-responsive drug delivery. Nanotechnology, 2017, 28: 275601.

[35] Zou Zhen, Li Siqi, He Dinggeng, et al. A versatile stimulus-responsive metal-organic framework for size/morphology tunable hollow mesoporous silica and pH-triggered drug delivery. Journal of Materials Chemistry B, 2017, 5: 2126-2132.

[36] Zha Fengchao, Shi Min, Li Hui, et al. Biomimetic mineralization of lipase@MOF biocatalyst for ease of biodiesel synthesis: Structural insights into the catalytic behavior. Fuel, 2024, 357: 129854.

[37] Miao Yangbao, Zhong Qilong, Ren Hongxia. Engineering a thermostable biosensor based on biomimetic mineralization hrp@Fe-MOF for alzheimer's disease. Analytical and Bioanalytical Chemistry, 2022, 414: 8331-8339.

[38] Yan Shuangqian, Zeng Xuemei, Wang Yu, et al. Biomineralized bacteria: Biomineralization of bacteria by a metal-organic framework for therapeutic delivery. Advanced Healthcare Materials, 2020, 9: 2070036.

[39] Zheng Qizhen, Sheng Jinhan, Liu Ji, et al. Histidine-rich protein accelerates the biomineralization of zeolitic imidazolate frameworks for *in vivo* protein delivery. Biomacromolecules, 2023, 24: 5132-5141.

[40] Shi Lingxia, Wu Jun, Qiao Xinrui, et al. In situ biomimetic mineralization on ZIF-8 for smart drug delivery. ACS Biomaterials Science & Engineering, 2020, 6: 4595-4603.

[41] Qi Xiaoyue, Chen Qizhe, Chang Ziyong, et al. Breaking pore size limit of metal-organic frameworks: Bio-etched ZIF-8 for lactase immobilization and delivery *in vivo*. Nano Research, 2022, 15: 5646-5652.

[42] Geng Peng, Yu Nuo, Macharia D K, et al. MOF-derived cus@Cu-MOF nanocomposites for synergistic photothermal-chemodynamic-chemo therapy. Chemical Engineering Journal, 2022, 441: 135964.

[43] Yu Zhanqing, Mao Wenjia, Lin Zihan, et al. Synthesis of porous carbon by composing Co-MOF as a precursor for degrading antibiotics in the water. Journal of Molecular Structure, 2023, 1271: 134131.

[44] Wilson B K, Sinko P J, Prud'homme R K. Encapsulation and controlled release of a camptothecin prodrug from nanocarriers and microgels: Tuning release rate with nanocarrier excipient composition. Molecular Pharmaceutics, 2021, 18: 1093-1101.

[45] Catania R, Onion D, Russo E, et al. A mechanoresponsive nano-sized carrier achieves intracellular release of drug on external ultrasound stimulus. RSC Advances, 2022, 12: 16561-16569.

[46] Lin Wenxin, Hu Quan, Jiang Ke, et al. A porphyrin-based metal-organic framework as a pH-responsive drug carrier. Journal of Solid State Chemistry, 2016, 237: 307-312.

[47] Jiang Ke, Ni Weishu, Cao Xianying, et al. A nanosized anionic MOF with rich thiadiazole

groups for controlled oral drug delivery. Materials Today Bio, 2022, 13: 100180.

[48] Chu Huiyuan, Shen Jiwei, Wang Chaozhan, et al. Biodegradable iron-doped ZIF-8 based nanotherapeutic system with synergistic chemodynamic/photothermal/chemo-therapy. Colloids and Surfaces A: Physicochemical and Engineering Aspects, 2021, 628: 127388.

[49] Teplensky M H, Fantham M, Li Peng, et al. Temperature treatment of highly porous zirconium-containing metal-organic frameworks extends drug delivery release. Journal of the American Chemical Society, 2017, 139: 7522-7532.

[50] Zhang Fengming, Dong Hong, Zhang Xin, et al. Postsynthetic modification of ZIF-90 for potential targeted codelivery of two anticancer drugs. ACS Applied Materials & Interfaces, 2017, 9: 27332-27337.

[51] Alavijeh R K, Akhbari K. Biocompatible MIL-101(Fe) as a smart carrier with high loading potential and sustained release of curcumin. Inorganic Chemistry, 2020, 59: 3570-3578.

[52] Dong Junliang, Yu Yueyuan, Pei Yuxin, et al. pH-responsive aminotriazole doped metal organic frameworks nanoplatform enables self-boosting reactive oxygen species generation through regulating the activity of catalase for targeted Chemo/chemodynamic combination therapy. Journal of Colloid and Interface Science, 2022, 607: 1651-1660.

[53] Peng Haibao, Zhang Xingcai, Yang Peng, et al. Defect self-assembly of metal-organic framework triggers ferroptosis to overcome resistance. Bioactive Materials, 2023,19: 1-11.

[54] Soomro N A, Wu Qiao, Amur S A, et al. Natural drug physcion encapsulated zeolitic imidazolate framework, and their application as antimicrobial agent. Colloids and Surfaces B: Biointerfaces, 2019, 182: 110364.

[55] Schnabel J, Ettlinger R, Bunzen H, et al. Zn-MOF-74 as pH-responsive drug-delivery system of arsenic trioxide. ChemNanoMat, 2020, 6: 1229-1236.

[56] Akbar M U, Khattak S, Khan M I, et al. A pH-responsive Bio-MIL-88b MOF coated with folic acid-conjugated chitosan as a promising nanocarrier for targeted drug delivery of 5-fluorouracil. Frontiers in Pharmacology, 2023, 14: 1265440.

[57] Liu Yuan, Gong C S, Dai Yunlu, et al. In situ polymerization on nanoscale metal-organic frameworks for enhanced physiological stability and stimulus-responsive intracellular drug delivery. Biomaterials, 2019, 218: 119365.

[58] Zhuang Jia, Kuo Chunhong, Chou Lienyang, et al. Optimized metal-organic-framework nanospheres for drug delivery: Evaluation of small-molecule encapsulation. ACS Nano, 2014, 8: 2812-2819.

[59] Alsaiari S K, Patil S, Alyami M, et al. Endosomal escape and delivery of Crispr/Cas9 genome editing machinery enabled by nanoscale zeolitic imidazolate framework. Journal of the American Chemical Society, 2018, 140: 143-146.

[60] Wang Haiyang, Wang Wenbo, Liu Lu, et al. Biodegradable hollow polydopamine@manganese dioxide as an oxygen self-supplied nanoplatform for boosting Chemo-photodynamic cancer therapy. ACS Applied Materials & Interfaces, 2021, 13: 57009-57022.

[61] Alkordi M H, Liu Yunling, Larsen R W, et al. Zeolite-like metal-organic frameworks as platforms for applications: On metalloporphyrin-based catalysts. Journal of the American Chemical Society, 2008, 130: 12639-12641.

[62] Sun Yi, Bao Bingbo, Zhu Yu, et al. An FPS-ZM1-encapsulated zeolitic imidazolate framework as a dual proangiogenic drug delivery system for diabetic wound healing. Nano Research, 2022, 15: 5216-5229.

[63] Rezaee R, Montazer M, Mianehro A, et al. Single-step synthesis and characterization of Zr-MOF onto wool fabric: Preparation of antibacterial wound dressing with high absorption capacity. Fibers and Polymers, 2022, 23: 404-412.

[64] Duan Fei, Wang June, Li Zhaoxi, et al. pH-responsive metal-organic framework-coated mesoporous silica nanoparticles for immunotherapy. ACS Applied Nano Materials, 2021, 4: 13398-13404.

[65] Chen Mian, Dong Ruihua, Zhang Jiangjiang, et al. Nanoscale metal-organic frameworks that are both fluorescent and hollow for self-indicating drug delivery. ACS Applied Materials & Interfaces, 2021, 13: 18554-18562.

[66] Cong Hailin, Jia Feifei, Wang Song, et al. Core-shell upconversion nanoparticle@metal-organic framework nanoprobes for targeting and drug delivery. Integrated Ferroelectrics, 2020, 206: 66-78.

[67] Liu Bei, Liu Zechao, Lu Xijian, et al. Controllable growth of drug-encapsulated metal-organic framework (MOF) on porphyrinic MOF for Pdt/Chemo-combined therapy. Materials & Design, 2023,.228: 111861.

[68] Doonan C J, Morris W, Furukawa H, et al. Isoreticular metalation of metal-organic frameworks. Journal of the American Chemical Society, 2009, 131: 9492-9493.

[69] Sun Nianxia, Zhang Min, Zhu Wentao, et al. Allyl isothiocyanate dry powder inhaler based on cyclodextrin-metal organic frameworks for pulmonary delivery. iScience, 2023, 26: 105910.

[70] Ma Peihua, Zhang Jinglin, Liu Ping, et al. Computer-assisted design for stable and porous metal-organic framework (MOF) as a carrier for curcumin delivery. LWT, 2020, 120: 108949.

[71] Lee D N, Gwon K, Kim Y, et al. Immobilization of antibacterial copper metal-organic framework containing glutarate and 1,2-Bis(4-pyridyl)ethylene ligands on polydimethylsiloxane and Its low cytotoxicity. Journal of Industrial and Engineering Chemistry, 2021, 102: 135-145.

[72] Shan Yongpan, Cao Lidong, Muhammad B, et al. Iron-based porous metal-organic frameworks with crop nutritional function as carriers for controlled fungicide release. Journal of Colloid

and Interface Science, 2020, 56: 383-393.

[73] Alijani H, Noori A, Faridi N, et al. Aptamer-functionalized Fe_3O_4@MOF nanocarrier for targeted drug delivery and fluorescence imaging of the triple-negative Mda-Mb-231 breast cancer cells. Journal of Solid State Chemistry, 2020, 292: 121680.

[74] Baati T, Njim L, Fadoua N, et al. In depth analysis of the *in vivo* toxicity of nanoparticles of porous iron(III) metal-organic frameworks. Chemical Science, 2013, 4: 1597-1607.

[75] Li Youcong, Su Jian, Zhao Yue, et al. Dynamic bond-directed synthesis of stable mesoporous metal-organic frameworks under room temperature. Journal of the American Chemical Society, 2023, 145: 10227-10235.

[76] Ruyra A, Yazdi A, Espín J, et al. Synthesis, culture medium stability, and *in vitro* and *in vivo* zebrafish embryo toxicity of metal-organic framework nanoparticles. Chemistry-A European Journal, 2015, 21: 2508-2518.

[77] Wuttke S, Zimpel A, Bein T, et al. Nanosafety: Validating metal-organic framework nanoparticles for their nanosafety in diverse biomedical applications. Advanced Healthcare Materials, 2017, 6: 1600818.

[78] Ledari R T, Shokat S Z, Qazi F S, et al. A mesoporous magnetic Fe_3O_4/BioMOF-13 with a core/shell nanostructure for targeted delivery of doxorubicin to breast cancer cells. ACS Applied Materials & Interfaces, 2023, 12: 14363.

[79] Zhao Xiaoyuan, He Shipeng, Li Bo, et al. Ducnp@Mn-MOF/Foe as a highly selective and bioavailable drug delivery system for synergistic combination cancer therapy. Nano Letters, 2023, 23: 863-871.

[80] Zhong Xiaofang, Zhang Yunting, Tan Lu, et al. An aluminum adjuvant-integrated nano-MOF as antigen delivery system to induce strong humoral and cellular immune responses. Journal of Controlled Release, 2019, 300: 81-92.

[81] Tan Guozhu, Zhong Yingtao, Yang Linlin, et al. A multifunctional MOF-based nanohybrid as injectable implant platform for drug synergistic oral cancer therapy. Chemical Engineering Journal, 2020, 390: 124446.

[82] de Lima H H C, da Silva C T P, Kupfer V L, et al. Synthesis of resilient hybrid hydrogels using Uio-66 MOFs and alginate (HydroMOFs) and their effect on mechanical and matter transport properties. Carbohydrate Polymers, 2021, 251: 116977.

[83] Pandey A, Kulkarni S, Vincent A P, et al. Hyaluronic acid-drug conjugate modified core-shell MOFs as pH responsive nanoplatform for multimodal therapy of glioblastoma. International Journal of Pharmaceutics, 2020, 588: 119735.

[84] Ahmed A, Karami A, Sabouni R, et al. pH and ultrasound dual-responsive drug delivery system based on peg-folate-functionalized iron-based metal-organic framework for targeted doxorubicin

delivery. Colloids and Surfaces A: Physicochemical and Engineering Aspects, 2021, 626: 127062.

[85] Li Zhen, Zhao Songjian, Wang Huizhen, et al. Functional groups influence and mechanism research of Uio-66-type metal-organic frameworks for ketoprofen delivery. Colloids and Surfaces B: Biointerfaces, 2019, 178: 1-7.

第 4 章 靶向与修饰

传统的药物递送系统通常会导致药物非特异性分布在全身，这可能导致靶位点的疗效降低、副作用增加以及生物利用度差。许多药物在生物环境中的溶解度或稳定性差从而导致生物利用度低，因此需要更高的剂量才能达到治疗效果。虽然药物可以快速代谢或从体内清除，但缩短其治疗窗口并需要频繁给药，这可能会影响患者的依从性。一些药物难以穿透更深的组织或穿过血脑屏障等生物屏障，从而限制了它们对某些疾病的有效性。

靶向递送系统可以通过将药物特异性地引导到疾病部位从而增加局部药物浓度，改善治疗效果，同时最大限度地减少全身暴露。靶向递送可减少对非靶组织的影响，从而减少与较高全身剂量相关的副作用和毒性。靶向系统通过改变药物释放速率和持续时间，可以在治疗窗口内更长时间地维持药物水平，从而降低给药频率。设计靶向系统来跨越生物障碍，增强药物向其他无法进入部位的递送，结合对特定生物条件（如 pH 值变化或酶的存在）做出反应的刺激反应材料可以在现场触发药物释放从而提高治疗效果。将 MOFs 整合到药物递送系统中，通过对其功能化，利用靶向/控释机制提高疗效，利用其高表面积和孔隙率实现有效的药物装载和释放。这种方法不仅解决了许多传统的缺陷，而且为先进且精确驱动的疗法开辟了新的途径。

4.1 功能化策略

用于药物递送应用的金属有机框架（MOFs）的功能化和表面改性是一个快速发展的研究领域。MOFs 具有高孔隙率和可调结构，为以可控方式封装和释放治疗剂提供了独特的优势[1]。然而，它们在生物医学中的直接应用往往受到生物相容性、生物环境中的稳定性和靶向递送能力等挑战的限制[2]。MOFs 材料的功能化策略包括表面改性、合成后修饰以及将活性位点或官能团掺入其结构中。这些策略增强了稳定性、生物相容性和靶向递送能力，这些特性适用于药物递送应用。MOFs 可以通过这些改性，针对特定应用进行定制，提高其有效性并扩大其潜在用途。

尽管一些 MOFs 材料因其自身存在的稳定性不足和相容性较差等缺陷，在应用上受到了一定限制，但表面修饰技术为解决这些问题提供了有效途径。通过表面修饰，不仅可以弥补 MOFs 材料的固有缺陷，还能赋予其新的功能，如增强靶向性并提供特定的结合位点。目前，MOFs 材料的修饰方法多种多样。其中，共价键外表面修饰是一种常见手段，例如使用聚乙醇类化合物进行修饰[3]，能够显著提升 MOFs 的稳定性与相容性。此外，外表面配位修饰也是一种有效的方法，它允许将多肽、脂质体、DNA 等[4]生物分子连接到 MOFs 的外表面，从而赋予其更广泛的生物应用潜力。除了通过化学键进行修饰外，还存在一些相对不稳定的修饰方法[5]，如包封和吸附等。虽然这些方法可能不如化学键修饰稳定，但它们为 MOFs 材料的应用提供了更多可能性，尤其是在需要灵活性和可逆性的场景中。MOFs 的独特特性，如高度有序的结构[6]、高表面积和大孔体积[7]，使其成为吸附和封装功能分子的理想平台。这些功能分子可以通过多种策略结合到 MOFs 中，包括“一锅法”、合成后修饰的共价键连接[8]等。这些策略使得功能化 MOFs 在治疗药物的生物应用中发挥重要作用，如表面吸附、孔隙封装以及共价结合等，为 MOFs 材料在生物医药领域的应用开辟了广阔前景。

4.1.1 非共价键结合

金属有机框架（MOFs）功能化中的表面吸附策略涉及治疗分子、催化剂或其他官能团直接附着在 MOFs 表面[9]。该方法被广泛用于提高 MOFs 的催化活性、选择性和吸附能力。通过将分子吸附到 MOFs 的外表面或孔隙内，可以针对特定药物输送过程定制框架的物理和化学特性。这种方法可以在不改变其固有结构的情况下相对直接地修改 MOFs。

例如通过表面吸附技术使得 MOFs 在酶固定化领域具有潜在应用价值[10]。早在 2006 年，Balkus 研究团队便报道了一项重要成果，他们成功将微过氧化物酶-11（MP-11）催化剂物理吸附在纳米晶铜基 MOFs 上，该催化剂在吸附后依然保持了其原有的催化活性。实验结果显示，Cu-MOFs 作为支撑材料，能够有效提升 MP-11 的催化性能。随后，Liu 等也开展了相关研究，他们合成了未经化学修饰表面的 MOFs，并将其用作酶-MOFs 生物反应器中的催化剂。研究深入探讨了主客体之间的相互作用机制，发现这种相互作用主要由氢键和π-π相互作用共同驱动。此外，Ma 等还探索了沸石咪唑酯框架（ZIFs）作为固定化酶基质的潜力。他们成功将亚甲基绿（MG）和葡萄糖脱氢酶（GDH）共固定在 ZIFs 上，制备出了一种集成式的电化学生物传感器[11]。在众多具有不同孔径、表面积和官能团的 ZIFs 中，ZIF-70 因其对 MG 和 GDH 出色的吸附能力脱颖而出。这些研究不仅展示了 MOFs 在酶固定化领域的广阔应用前景，也为后续研究者提供了宝贵的经验和启示。Zhou 小组[12]成功把三种不同的酶封装到 PCN-333（Al）制备了稳定的 PCN-333，包含大型介孔笼作为单分子陷阱（SMTs）用于酶封装，防止酶的聚集和浸出。

核酸也可以通过表面吸附的方式被有效地固定在 MOFs 上。以 Zhou 和 Deng 小组的研究为例，他们精心设计了四个具有等网状结构的 MOFs，其开放孔道的尺寸从 2.2 nm 逐渐增加到 4.2 nm，这种设计能够精确地容纳单链 DNA（ssDNA）分子[13]。MOFs 在这里扮演了一个出色的宿主角色，它通过将核酸链完全包裹在孔道内部，有效地保护了 ssDNA 免受降解。研究表明，MOFs 中通道大小的合适选择以及适度的调节，为核酸与框架之间提供了范德瓦耳斯相互作用。这种相互作用在核酸的可逆摄取和释放过程中起到了关键作用，使得 MOFs 成为一种理想的核酸固定化平台。这一研究不仅展示了 MOFs 在核酸固定化方面的潜力，还为后续研究者提供了有益的启示。随着技术的不断进步和研究的深入，相信 MOFs 在生物医药、生物传感等领域的应用将会更加广泛和深入。

抗癌方面，Yin 等[14]构建新型的 AuNC/Cam@MOF 多功能纳米探针，NH_2-MIL-101（Fe）通过非共价封装装载了大量的喜树碱（Cam），用于靶向传递抗癌药物和通过实时荧光成像和化疗。Yao 等[15]通过吸收光谱和密度泛函理论（DFT）计算，研究了阿霉素与纳米 MOFs 之间的相互作用机制，发现脱质子化的阿霉素被静电吸附在不饱和铁簇上。有研究将多功能 MOFs 纳米探针通过非共价封装方式，将喜树碱药物成功组装进 MOFs 中，并与叶酸作为目标识别元素进行结合。此外，利用标记的底物肽作为识别部分和信号开关，进一步增强了 MOFs 探针的功能性和精准性。这款精心设计的 MOFs 探针不仅能够与组织蛋白酶紧密结合，还能携带喜树碱药物，实现针对组织蛋白酶 B 激活的癌细胞的双重治疗作用。具体而言，它能够用于癌细胞成像，提供精准的诊断信息；同时，还能通过化学光动力学治疗杀灭癌细胞，达到治疗目的。相较于单独使用某一种治疗方法，这种双功能纳米探针在化学疗法所需的时间、激光功率以及光动力学治疗所需的照射时间等方面均展现出了更高的治疗效率[16]。这一突破性的设计为癌症治疗提供了新的策略和希望。

Soomro 等[17]通过一步法成功地将天然抗菌药物 PHY 封装到 ZIF-8 中，琼脂孔扩散法测 PHY@ZIF-8 对微生物具有最大的抗菌活性，因此该药物可能具有治疗由微生物引起的感染相关疾病的潜力。

将功能分子纳入 MOFs 的一种高效且多功能的策略是通过“一锅法”的方式实现封装孔[18]。这种方法巧妙地将 MOFs 的形成与底物的封装过程合二为一，从而使得相比 MOFs 的孔径，更大尺寸的分子也能被稳定地固定在 MOFs 的空腔内部。需要注意的是，这种封装技术要求底物在合成过程中必须能够保持稳定，以确保封装的成功[19]。以 Zhuang 等的研究为例，他们成功地将抗癌药物喜树碱封装在 ZIF-8 框架内，制备出了粒径均匀（约 70 nm）的单分散纳米球[20]。这种药物封装的 MOFs 纳米球在针对 MCF-7 乳腺癌细胞系的研究中表现出了增强的细胞内化效果和降低的细胞毒性，展示了其在癌症治疗中的潜在应用价值。另外，Ding 等[21]也通过巧妙地混合无机金属盐、有机配体和药物分子，成功地将抗癌药物 3-甲基腺嘌呤纳入 ZIF-8 中。实验结果显示，在用 3-甲基腺嘌呤@ZIF-8 纳米颗粒处理的 HeLa 细胞中，自噬抑制的强度得到了显著增加，进一步证明了这种封装策略在药物传递和疗效增强方面的有效性。总的来说，ZIF-8 由于其优异的单分散性、适宜细胞摄取的尺寸、温和的合成条件以及易于进行表面修饰的特点，被认为是一种理想的细胞内药物传递宿主材料。通过从头合成的方式实现封装孔，可以将多种功能分子有效地纳入 MOFs 中，为药物传递和癌症治疗等领域提供了新的思路和策略。

4.1.2 共价键结合

共价键和功能分子封装策略涉及通过化学键或使用功能分子作为构建单元将治疗剂整合到用于药物递送的 MOFs 结构中。这确保了药物的稳定掺入，从而可以精确控制释放速率。通过将药物与 MOFs 共价连接或构建以药物分子为成分的 MOFs，可以实现增强的稳定性和靶向递送，从而改善治疗结果并减少副作用。在 MOFs 表面的共价变化方面已进行了许多类型的研究，例如，“点击”化学[22]、质子化[23]、烷基化[24]、还原[25]、亚胺缩合[26]、溴化[27]和酰胺偶联[28]。一般而言，MOFs 表面富含多种官能团，诸如氨基、羧基以及羟基等，这些官能团可以与目标分子上的活性基团有效形成共价键[29]，从而实现精准的结合与调控。这一特性使得 MOFs 在药物传递、催化以及生物传感等领域具有广阔的应用前景。通过合理设计和调控 MOFs 表面的官能团种类和数量，可以实现对目标分子的高效捕获与释放，为相关领域的研究和应用提供有力支持。

Yan 等研究表明[30]，CAL-B-MOF 生物偶联物对(±)-1-苯乙醇的对映选择性和活性得到了很好的保持，这证明了该偶联策略的有效性和可靠性。进一步地，他们采用了类似的偶联方法，成功地将蛋白酶特别是胰蛋白酶固定在 MIL-88B（Cr）、MIL-88B-NH_2（Cr）和 MIL-101（Cr）这三种 MOFs 上[31]。这一过程主要是通过胰蛋白酶的胺基团对 DCC 激活的 MOFs 进行亲核攻击来实现的，从而实现了酶与 MOFs 之间的牢固结合。值得注意的是，胰蛋白酶与 MIL-88B-NH_2（Cr）的结合物展现出了与天然胰蛋白酶相当的牛血清白蛋白（BSA）蛋白消化能力。这一结果不仅证明了固定化后的胰蛋白酶仍保持了其原有的活性，也展示了 MOFs 作为酶固定化载体的潜力。除了羧酸基外，他们还探索了利用氨基上的有机配体来与酶进行配合。例如，葡萄糖氧化酶[32]和大豆环氧化物水解酶等酶类也可以通过这种策略与 MOFs 进行有效结合（图 4.1）。这种方法不仅拓宽了 MOFs 在酶固定化领域的应用范围，也为开发新型、高效的生物催化剂提供了新的思路。通过合理的偶联策略和选择适当的 MOFs 材料，可以实现酶的高效固定化并保持其原有的活性。这将有助于推动生物催化、药物传递和生物传感等领域的发展，为未来的科学研究和技术应用提供有力支持。

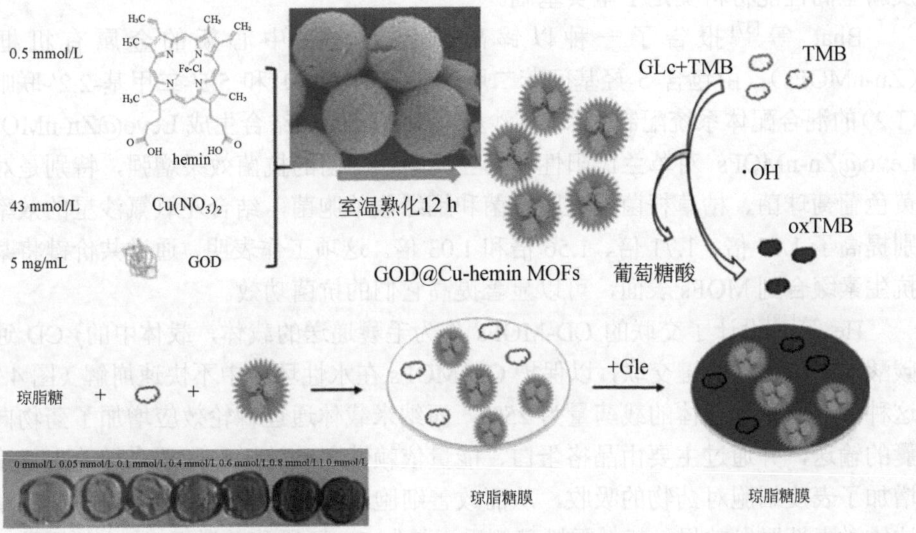

图 4.1　基于 Cu-MOFs 的抗菌研究[32]

在有机连接物上的点击反应已被用于固定化生物分子[33]。Mirkin 小组报道了核酸-MOFs 纳米颗粒偶联[34]的第一个例子。它们是通过叠氮化物功能化的 UiO-66 和二苯基环辛基功能化的 DNA 之间的促进“点击”反应合成的。由于 UiO-66 的孔径较小，因此 DNA 链与 MOFs 纳米颗粒的外表面相协调。UiO-66 的结构在化学反应过程中可以保持不变。与非功能化的 MOFs 纳米颗粒相比，DNA-MOFs 偶联物表现出更高的胶体稳定性和增强的细胞摄取。

除了有机连接物，无机金属团簇在 MOFs 中也扮演着重要角色，它们提供了另一种反应位点，使得功能分子能够与之共价结合。在 2017 年，Mirkin 研究小组[35]报道了一种创新的方法，该方法能够直接且普遍地将寡核苷酸功能化到 MOFs 纳米颗粒的外表面。他们采用了一种基于配位化学的策略，成功地将 MOFs 纳米颗粒的外部金属节点与末端磷酸盐修饰的寡核苷酸实现共价连接。这种策略的成功应用使得 9 种具有不同金属（如 Zr、Cr、Fe 和 Al）的典型 MOFs 均能够在外表面被寡核苷酸有效修饰。这种方法的优势在于，它能够使粒子的功能化表面独立于 MOFs 的整体结构，从而提供了更大的灵活性和可控性。此外，DNA 的化学可编程性使得我们能够精确操纵粒子间的相互作用，进一步拓宽了 MOFs 在材料科学和生物医学等领域的应用前景。另一项值得关注的研究来自 Zhou 小组，他们采取“一锅法”[36]成功地将一系列卟啉衍生物纳入到稳定的 Zr-MOFs 中。这种方法利用了 Zr 簇上现有的配位位点，通过混合不同几何形状和连接性的配体，成功地将四聚四酮（4-羧苯）卟啉（TCPP）配体整合到 MOFs 中，同时保持了原始 MOFs 的晶体结构。这一研究不仅为 MOFs 的功能化提供了新的策略，也为开发新型高性能材料奠定了坚实基础。

Bhat 等[37]报告了一种以锌离子作为金属中心新的金属有机框架（Zn-nMOFs），由包含 5-羟基间苯二甲酸酯离子（L1）和 5,5′-二甲基-2,2′-联吡啶（L2）的混合配体系统配制。左氧氟沙星与 MOFs 成功结合生成 Levo@Zn-nMOFs。Levo@Zn-nMOFs 对革兰氏阳性和革兰氏阴性细菌的抗菌效果增强，特别是对金黄色葡萄球菌、枯草杆菌、大肠杆菌和铜绿假单胞菌，结合左氧氟沙星的效率分别提高了 1.65 倍、1.71 倍、1.56 倍和 1.03 倍。这项工作表明，通过共价键将某些抗生素缀合到 MOFs 表面，可以显著提高它们的抗菌功效。

He 等[38]设计了交联的 CD-MOFs 作为毛囊递送的载体，载体中的γ-CD 通过碳酸二苯酯以共价键交联，以保护 CD-MOFs 在水性环境中不快速崩解（图 4.2）。这种改进的纳米载体的载药量为 25%，而纳米载体通过棘轮效应增加了药物向毛囊的输送，并通过主要由晶格蛋白、能量依赖性主动转运和脂质介导的内吞途径增加了表皮细胞对药物的吸收，从而改善细胞活力、增殖和迁移，进而显著增强抗雄激素性脱发效果，其中雪松醇侧重于抑制 5α-还原酶并激活 Shh/Gli 通路，而米诺地尔则上调 VEGF，下调-调节 TGF-β，并激活 ERK/AKT 途径。

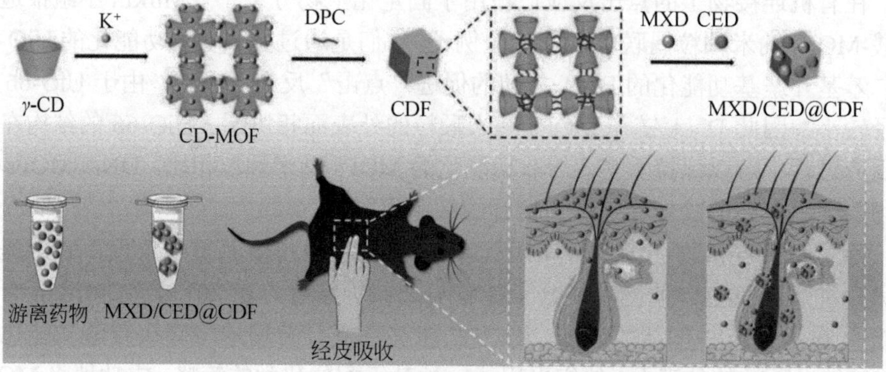

图 4.2　CD-MOFs 作为毛囊递送的载体[38]

4.1.3　功能性分子作为构建模块

另一种功能化 MOFs 的有效方法是通过设计功能分子并将其作为构建模块来实现。生物分子因其独特的化学性质，通常含有多种可以与无机金属进行配位的活性化学基团，因此在功能化 MOFs 中发挥着重要作用。例如，氨基酸[39]、多肽[40]、碱基以及糖类[41]等生物分子已被广泛用作有机配体来合成 Bio-MOFs。这类 Bio-MOFs 不仅具备良好的生物相容性，还具备特殊的生物功能，为生物医学领域的应用提供了广阔的前景。然而，值得注意的是，大多数生物分子因其高度的灵活性和较低的对称性，使得直接利用它们来形成高质量的 MOFs 晶体成为一项具有挑战性的任务。为了克服这一难题，研究者们需要对生物分子进行精细的设计和调控，以确保它们能够与金属离子有效地进行配位，进而形成稳定且有序的 MOFs 结构。

为了形成一个以低对称腺嘌呤为构件的高度有序的 MOFs 结构，有研究[42]将双苯基二羧酸（BPDC）、腺嘌呤和乙酸锌混合引入了一种对称的共配体来指导合成了结晶多孔生物-MOF-1。MOFs 由无限的一维锌腺苷酸柱组成，它们通过线性 BPDC 连接器相互连接，由角融合的锌腺苷酸八面体构建单元（ZABUs）组成。随后，有小组[43]报道了一种介孔 MOFs，bio-MOF-100，具有较高的表面积（4300 m^2/g）和孔隙体积（4.3 cm^3/g）。通过 MOFs 材料独特的桥连方式，生成一个具有大空腔和孔道的三维结构。MOFs 的药物传递系统通常是通过药物分子在材料孔内的非共价吸附或通过有机连接基的共价修饰而构建的。据报道，Pander 提出了使用溶剂辅助配体结合将药物分子与金属节点结合的方法，从而将药物分子结合到 MOFs 中。他们测试了 NU-1000 平台，该平台包含 8 个连接的 Zr 节点，该节点最多可以结合 4 个基于羧酸盐的配体。成功地锚定了三个含有羧酸盐基团的模型药物分子，即酮洛芬、萘啶酸和左氧氟沙星，且药物载量达到 30%（质量分数）[44]。

4.2　外源性刺激响应药物靶向递送系统的设计

外源性刺激响应药物递送系统的一个显著优势在于其能够实现对刺激时间、位置和强度的精准控制。这种精准调控不仅使得外部刺激的管理更为完善，更能有效地减少对正常组织和器官的潜在损伤，从而实现治疗的安全性和精准性。更为出色的是，外部刺激响应系统还具备刺激之间相互转换的能力。这意味着在复杂的医疗场景中，系统可以灵活地应对不同的刺激需求，根据实际需要切换或组合不同的刺激方式，以达到最佳的治疗效果。这种灵活性和适应性使得外源性刺激响应药物递送系统在医疗领域具有广阔的应用前景。基于 MOFs 的外源性刺激响应递送平台可以分为以下四类。

4.2.1　光响应药物递送平台

近年来，近红外（NIR）光因其在安全性和组织穿透性方面的卓越表现，被越来越多地应用于药物释放[45]的控制。与紫外光或白光相比，NIR 的这些独特优势使其成为了一个更为理想的选择。Xu 等研究并介绍了一种对 NIR 敏感的茶多酚改性锆基卟啉金属有机框架（TP-Au@PCN）来作为治疗骨性关节炎（OA）的多功能纳米平台[46]。分子生物学研究表明，TP-Au@PCN 可降低软骨细胞凋亡，增加软骨细胞代谢 80.7%。软骨细胞表达的Ⅱ型胶原蛋白和蛋白多糖含量随后增加。注射 TP-Au@PCN 可有效地将关节温度提高到 46.9℃，并在 NIR 激发下控制给药。Sheng 等[47]报告了一种基于 MOFs 的蛋白质递送系统，该系统能够通过深层组织可穿透的近红外（NIR）光精确控制蛋白质释放。通过将目标蛋白和上转换纳米粒子（UCNPs）封装在沸石咪唑酯框架 MOFs（ZIF-8）中，并进一步将光酸生成剂（PAG）捕获在 ZIF-8 的孔隙中，以此构建响应系统。在 NIR 光照射下，UCNPs 发出紫外光，激活 PAG 产生质子，这使得局部酸化，从而使 ZIF-8 降解以实现时空控制的蛋白质释放。

1. 光热响应系统

部分 MOFs 材料本身便展现出卓越的光热转换能力，而通过对其进行改性处理，可以进一步增强其抗菌性能，使之成为高效抗菌剂[48]的理想制备材料。以普鲁士蓝掺杂 Zn 离子为例，改性后的颗粒不仅保持了普鲁士蓝原有的空间结构，而且还拥有了更加稳定[49]的结构特性，为其在抗菌领域的应用提供了坚实基础。有研究报道了一种在低 pH 值时，由于壳体的损伤，释放的阿霉素不断增加的 MOFs[50]。当阿霉素负载的 PPy@MIL-100（Fe）NCs 暴露于 NIR 照射下时，局部温度升高，导致阿霉素释放速度更快。采用 PPy@MIL-100（Fe）对癌细胞进行双模式 MRI/PAI 和协同化疗-光热治疗。

2. 光动力学响应系统

金属有机框架（MOFs）的光动力学响应平台原理基于 MOFs 能够作为荧光材料和光敏剂载体的特性，利用光动力学治疗（PDT）进行癌症治疗或环境传感。光动力学治疗是一种利用光激活光敏剂来产生活性氧（如单线态氧和含氧自由基），这些活性氧可以杀死癌细胞、病原体或引起特定化学反应的治疗方法。

在光动力学治疗（PDT）过程中，由于能够产生 ROS（活性氧）并导致肿瘤组织内部氧气消耗，我们可以利用这一特性，将 ROS 响应位点或缺氧响应位点整合到纳米材料中，制备出具有光动力学响应性的纳米载体用于药物递送。这样的设计使得药物分子能够在肿瘤组织内实现选择性释放[51]，从而提高治疗的精准性和有效性。此外，二硒键（—Se—Se—）作为一种特殊的化学键，能够被 ROS 氧化为硒醇或硒酸，因此也常被应用于基于光驱动的 ROS 响应型药物[52]递送系统中，为精准医疗领域的发展提供了新的思路和方法。Yang 等研究构建了一种光响应触发具有类过氧化氢酶活性和 GSH 过氧化物酶样活性的过氧化氢响应的纳米结构 $MnFe_2O_4$@MOF@PEG[53]用于连续自生成 O_2，同时降低 TME 中的谷胱甘肽

（GSH）能力，从而增强 PDT 的抗肿瘤治疗效果。

3.光转化型响应系统

此外，尽管众多光敏性官能团，如香豆素和偶氮苯等，在紫外光或短波长蓝光的照射下能够发挥其独特作用[54]，然而这些光源在生物体应用时受到组织穿透力不足的严重制约，极大地限制了它们在生物体内的应用前景。为了克服这一难题，研究人员进行了深入的探索，他们尝试将近红外光转换为短波长光[55]，并成功将具有这种转换功能的纳米颗粒引入到纳米药物递送系统中。这种创新的设计使得纳米药物递送系统能够在细胞内或组织内发光，从而有效地解决了光穿透力不足的缺点。尽管这一策略在药物递送领域展现出了巨大的应用潜力，但由于技术实现上的难度，目前仍面临着一定的挑战。

4.2.2 磁响应药物递送平台

MOFs 由于其独特的结构和性质，使得研究人员可以通过设计引入磁性中心，从而赋予 MOFs 磁响应特性[56]。这种磁性中心通常是由某些具有磁性的金属离子或金属簇构成的。磁性药物输送作为一种有效的方法，通过结合涂有载药基质的易磁性材料，将药物直接运送到身体的器官或目标位置。MOFs 的磁性纳米粒子作为药物载体，通过靶向治疗位置而不影响其他细胞。磁辅助将治疗剂递送到目标部位，这被称为磁性药物靶向，在许多研究中已被证明是一种有前途的策略。

其相对于其他靶向策略的主要优势之一是可以在外给药后远程控制纳米载体的分布和积累。当这些具有磁响应功能的 MOFs 暴露在外磁场下时，它们会受到磁场的吸引或排斥作用，从而实现定向移动或定位[57]。这种磁响应性使得 MOFs 在生物医学应用中具有巨大的潜力，例如在靶向药物输送[58]、磁共振成像[59]等领域。磁响应功能的实现需要考虑到 MOFs 的稳定性、生物相容性以及磁响应强度等因素。因此，在设计和合成具有磁响应功能的 MOFs 时，需要综合考虑这些因素，以确保其在实际应用中能够发挥最佳效果。

在靶向药物输送方面，研究人员可以将药物分子负载在具有磁响应功能的 MOFs 上，并通过外部磁场引导这些 MOFs 到达特定的病变部位[60]。一旦 MOFs 到达目标位置，药物分子就可以被释放出来，实现对疾病的精准治疗。Din 等[61]报道了通过将超顺磁性γ-Fe_2O_3 NPs 整合到 MOF MIL-53（Al）中，通过一步原位热解途径合成γ-Fe_2O_3@MIL-53（Al）。

多孔磁性核-壳复合物[62]因其独特性质在众多领域中得到了广泛的应用，其魅力在于结合了壳的多孔性与核的磁学性两大优势。鉴于 MOFs 材料所展现的其他孔材料无法比拟的独特性质，基于 MOFs 的多孔磁性核-壳结构的设计与合成显得尤为合理且具有重要意义。这种磁性复合物在可循环利用催化[63]、磁性分离[64]、靶向载药[65]以及磁共振成像[66]等多个领域均展现出潜在的应用前景，预示着它将在未来为这些领域带来革新与突破。Akbar 等[58]由预组装的 Fe_2Mn（μ_3-O）簇合成了基于 Fe、Mn 的铁磁 MOFs。Mn 的引入为 FeMn-MIL-88B 提供了铁磁特性。5-氟尿嘧啶（5-FU）作为模型药物封装在 MOFs 中，实现了 pH 值和磁性双刺激响应控释。FeMn-MIL-88B 在肿瘤微环境（TME）模拟介质中具有 43.8%的更高 5-FU 负载能力和快速药物释放行为。Li 等[67]使用 Fe^{3+}和 2-氨基-1,4-苯二甲酸酯（BDC-NH_2）作为磁性金属有机框架（MOFs，Fe_3O_4-NH_2@MIL101-NH_2）核壳纳米粒子，通过微波辐射在水相中快速合成金属离子和配体。所得磁性 MOFs 具有大表面积（96.04 m^2/g）、优异的磁响应（20.47 emu/g）和大介孔体积（22.07 cm^3/g）以及直径范围为 140～330 nm 的球形形态（图 4.3）。

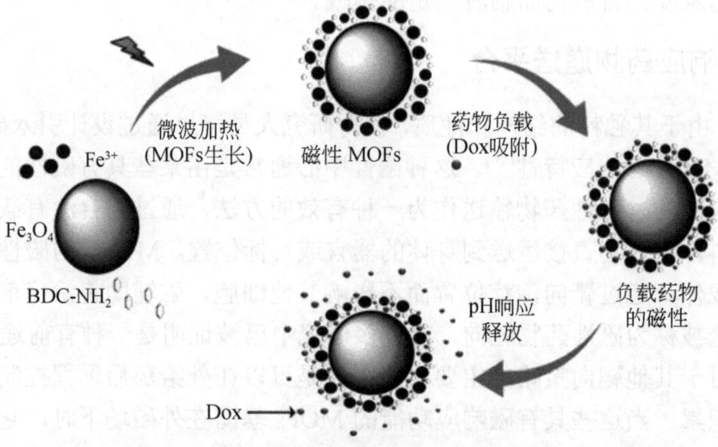

图 4.3　磁响应 MOFs 装载阿霉素（Dox）[67]

最近，Bhattacharjee 等[68]开发了一种基于 MIL-100（Fe）的 DDS，负载抗癌药物盐酸阿霉素（Dox）；MIL-100 的负载能力明显提高。另有，Yang 等制作了一个辐射 MOFs 壳和一个 pH 敏感羟基磷灰石（HAP）的一种新的多孔复合材料 Fe_3O_4[69]。将抗癌药物 Dox 成功地负载到 Fe_3O_4@Fe-MOF@HAP 纳米材料上，发现其承载能力为 75.38 mg/g。另一个磁性多孔 MOFs 的例子由 Bellusci 等报道，他们开发了研磨方法来制备具有磁性和高孔隙率的 MOFs[70]。柯飞等[71]也提出了一种简便且可控的方法，旨在设计和合成新型的磁性核-壳 MOFs 微球 Fe_3O_4@MOF。此外，他们还深入探究了这种磁性微球在多个领域中的潜在应用价值，为其在实际应用中的广泛推广奠定了坚实的基础。Chung 等[72]开发一种具有双重释放阶段（即爆发和稳定释放）的磁响应释放材料，用于选择性激活相关生理学和治疗细菌感染的皮肤伤口。

4.2.3　热响应药物递送平台

热响应平台在 MOFs 中的作用主要依赖于 MOFs 材料对温度变化的敏感响应[73]。这种敏感响应通常是通过 MOFs 中某些组分的热敏性质来实现的，这些组分可以在温度变化时发生物理或化学变化，进而改变 MOFs 的整体性质或功能。热响应平台的作用机制可以涉及以下几个方面：MOFs 中的有机配体或金属离子可能在特定温度下发生构型转变，导致 MOFs 的孔道结构或孔径大小发生变化[74]。这种变化可以影响 MOFs 对气体或分子的吸附能力，从而实现热响应的吸附/脱附过程。

某些 MOFs 材料在特定温度下会发生相变，从一种晶体结构转变为另一种晶体结构[75]。这种相变可以改变 MOFs 的物理和化学性质，如导电性、光学性质等，从而实现热响应的功能。MOFs 中常常含有溶剂分子，这些溶剂分子在温度变化时可能发生脱附或重新吸附。通过选择合适的溶剂和 MOFs 结构可以调控这种溶剂交换过程，使 MOFs 在特定温度下表现出不同的性质或功能。当 MOFs 被用作药物输送载体时，通过热响应平台可以实现药物在特定温度下的释放。例如，当 MOFs 受到外部热源加热时，其结构中的药物分子可能因热运动而加速释放，从而实现靶向给药[76]。多孔 MOF 可以通过物理吸附法将药物吸附在孔隙表面，形成主客体相互作用[77]。二级键[78]、偶极子[79]、电荷或氢键[80]，在主客体相互作用中不如化学键强。分子识别的熵代价导致主客体相互作用在加热时失去稳定性[81]。因此，在高温下减弱的主客体相互作用可以驱动快速释放速率。

有研究设计了一种锆基金属有机框架 ZJU-801 载药系统[82]，其载药量高达 41.7%。ZJU-801 与双氯芬酸钠（DS）之间配体体积大、稳定性好、π-π相互作用强烈，从而成功实现了指令加热激活药物释放。Wang 等制备了基于镧系多金属氧酸盐的发光温度/pH 双响应纳米颗粒，作为一种可控的药物递送系统[83]。在药物递送载体中，介孔二氧化硅作为核心，聚（*N*-异丙基丙烯酰胺–邻甲基丙烯酸）和镧系多金属氧酸盐分别作为双响应外壳和发光标记，这种复合水凝胶在紫外线照射下表现出肉眼可见的红色发光[84]。盐酸阿霉素从复合物中的释放是温度/pH 依赖性[85]。Nannuri 等[86]将碳点（CDs）集成到 MOFs 中将进一步增强其功能和成像能力。在这项研究中，载有碳点的 MOFs 的合成和表征被详细描述为一种用于神经母细胞瘤成像和治疗的新型 pH 和温度响应纳米平台。使用改进的微波方法合成碳点。碳点在 405 nm 激发下也表现出绿色发射具有浓度依赖性发射行为。据观察，随着温度升高（30℃至 65℃），以 5℃的步长覆盖生理温度范围，荧光强度降低了约 50%。

通过控制环境条件从载体材料中控制小分子的释放具有用于药物递送的潜力。Nagata 等[87]报道了既具有 pH 响应性又具有热响应性 MOFs 载药系统，允许客体分子以“开-关”方式从 MOFs 释放出来。在高 pH（6.86）或低温（＜25℃）下，当聚合物采用线圈构象时，客体分子会从 MOFs 中快速释放，而在低 pH（4.01）或高温（＞40℃）下，当聚合物采用球状构象时，客体分子的释放受到抑制。即使开始释放后，也可以通过施加外部刺激来停止释放。双重调控有助于药物的更精准释放，Zhang 研究用 *N*-异丙基丙烯酰胺（NIPAM）和丙烯酸（AA）对 MOFs 进行表面改性装载普鲁卡因胺，合成的 P（NIPAM-AA）是由双重外部刺激触发的受控释放的典型例子[88]。通过调节 pH 值和温度来实现两种开关状态之间的轻松切换。

4.2.4　声响应药物递送平台

MOFs 的声响应平台主要利用声波与 MOFs 材料之间的相互作用来实现特定的功能[89]。声波作为一种机械波，可以通过振动传递能量。当声波作用于 MOFs 材料时，这种能量传递可以引起 MOFs 内部结构的动态变化，从而实现声响应。

声波可驱动 MOFs 的结构变化，在特定频率和强度的声波作用下，MOFs 的框架结构可能会发生可逆的形变或振动[90]。这种结构变化可以影响 MOFs 的孔道大小和形状，进而调控其对气体或分子的吸附和释放能力。声波能量可以激活 MOFs 中的某些化学反应，如催化反应或化学键的断裂与形成[91]。这种声化学效应使得 MOFs 在声波作用下能够表现出特殊的催化性能或响应性[92]。当 MOFs 以纳米粒子或微粒的形式存在时，声波可以通过声流效应或声辐射力驱动这些粒子的运动[93]。这种运动可以实现对 MOFs 粒子的定向操控、聚集或分散，从而在药物输送、传感或环境治理等领域实现特定的应用。需要注意的是，MOFs 的声响应平台作用机制相对复杂，且受到多种因素的影响，如声波频率、强度、作用时间以及 MOFs 的结构和性质等。因此，在设计和应用具有声响应功能的 MOFs 时，需要进行深入的研究和优化，以确保其在实际应用中能够发挥最佳效果。

有研究设计合成了 PEG-FA-NH_2-Fe-BDC，280 min 的高超声触发释放效率为 90%，证实了该 NPs 的超声敏感性[94]。通过细胞活力测定证实了在体内使用的浓度下的低细胞毒性，同时通过流式细胞术证实了较高的药物内在化和性能[95]。目前，关于 MOFs 声响应平台的研究仍处于初级阶段，但其在药物输送[96]、环境监测[97]、生物医学成像[98]等领域具有潜在的应用前景。随着科研技术的不断进步和应用需求的不断增长，相信未来会有更多关于 MOFs 声响应平台的研究成果涌现。

4.3　内源性刺激响应药物靶向递送系统的设计

4.3.1　氧化还原响应纳米药物递送平台

1. 还原型响应

由于肿瘤组织中谷胱甘肽（GSH）水平的升高，肿瘤细胞内呈现出一种还原性环境。这种特殊的细胞环境不仅调节了细胞内的代谢过程，还成为了向肿瘤细胞递送药物的独特切入点[99]。与细胞外环境相比，细胞内的 GSH 浓度更高，为利用氧化还原反应设计药物递送系统提供了可能性。基于氧化还原反应的纳米载体因其多重优势而受到广泛关注。它们具有优异的稳定性、高药物负载能力、对亲脂性药物的良好递送能力[100]以及快速的药物释放能力。这些特性使得氧化还原响应纳米载体在药物递送领域具有广阔的应用前景。在氧化还原反应物质中，二硫键（—S—S—）[101]、二硒键（—Se—Se—）[102]以及某些高价金属离子等扮演着重要角色。这些物质大多被应用于药物递送系统中，以实现药物的靶向释放。在 GSH 的作用下，二硫键和二硒键能够发生交换反应，从而被 GSH 破坏断裂。同样地，高价金属离子在 GSH 的作用下也能被还原为低价金属离子。这些氧化还原反应不仅确保了药物递送系统的高效性，还为肿瘤治疗提供了新的策略和方法。

在设计针对谷胱甘肽（GSH）响应的纳米颗粒时，二硫键在生物硫醇作用下的裂解[103]是一个被广泛采用的策略。为实现这一目的，研究者们通常会在聚合物链中引入二硫键，用以连接前药物分子或作为药物传递的载体。这些经过巧妙设计的两亲性聚合物能够自组装形成纳米颗粒，并通过增强渗透和滞留（EPR）效应[104]在肿瘤部位实现有效富集。当纳米颗粒进入肿瘤细胞后，在 GSH 的作用下，二硫键会发生断裂，从而触发药物的释放。这一策略不仅提高了药物在肿瘤部位的选择性聚集，还实现了药物的精确释放，为肿瘤治疗提供了更加精准和高效的方法。

多价态金属离子常常展现出与谷胱甘肽（GSH）反应的特性。当这些金属离子处于高价态时，它们可以在 GSH 的作用下被还原为相应的低价态[105]。在常见的金属离子中，具有多价态组合的如铁（Fe）、锰（Mn）和铜（Cu）等都表现出了这样的性质。这种与 GSH 的反应性质使得多价态金属离子在药物递送和肿瘤治疗等领域具有潜在的应用价值。通过利用这些金属离子的价态变化，我们可以设计出更为精准和有效的药物递送系统，以实现对肿瘤的高效治疗。近期 Zhang 和他的同事报道了一种生物相容的多功能基于 MOF 的 DDS，具有 pH 和 GSH 双响应性和良好的降解性，通过绿色“一锅法”表面修饰策略递送抗癌药物阿霉素[106]。

肿瘤细胞内的还原性物质——谷胱甘肽（GSH）的浓度显著高于细胞外环境，达到了 10 倍以上的水平，同时相较于正常细胞，其浓度也高出 4 倍之多。这种高浓度的 GSH 在药物递送过程中发挥着关键作用，它可以触发药物分子中的二硫键发生断裂，进而在细胞质中实现药物的精准释放[107]。基于这一发现，他们提出了一种创新的药物载体设计思路：将含巯基键的咪唑与 Mn 结合，用于装载化疗药物顺铂（CDDP）。通过纳米工程改造技术，将这种复合物与具有同源靶向性的相变材料（PCM）和肿瘤细胞膜（TCM）相结合，成功制备出 Mn-MOF-CDDP@PCMTCM 纳米导弹[108]。这种纳米导弹通过主动和被动靶向机制，能够精准地定位于肿瘤细胞。在微波热敏化和谷胱甘肽的双重响应下，纳米导弹能够在肿瘤细胞内定点释放 CDDP 药物，从而有效逆转 CDDP 的耐药性，并显著增强微波热动力化疗的治疗效果。体内外实验结果表明，该纳米导弹对耐药肿瘤模型具有优异的治疗效果，为肿瘤治疗领域提供了新的希望和方向。

有研究开发了一种基于 Mn^{2+}和二硫代二乙醇酸的新型氧化还原敏感 MOFs 材料[109]，GSH 增敏剂诱导机制通常是利用金属离子（如 Mn(Ⅱ)[110]和 Cu(Ⅱ)[111]等）的氧化还原活性来释放负载药物。Wan 等构建了 GSH 锁定 Mn(Ⅲ)密封 MOFs 纳米系统，通过可控活性氧（ROS）的产生和 GSH 消耗[112]，实现了监测光动力学治疗（PDT）的双模式。肿瘤细胞内吞后获得的纳米平台可以与细胞内 GSH 反应分解，释放 Mn(Ⅱ)和游离 TCPP，最终激活基于 Mn(Ⅱ)的磁共振成像（MRI）和基于 TCPP 的光学成像（OI）。肿瘤内注射 MOFs 研究了 GSH 对其反应的影像学效果和 MRI 表现。从 0 至 30 min，肿瘤部位的荧光显著增加，表明体内 GSH 逐渐从 MOFs 中释放 TCPP，导致 GSH 解锁近红外荧光。

2. 氧化型响应

癌细胞中 GSH 的过表达，实际上是细胞对于内部高水平活性氧（ROS）的一种代偿反应。高水平的 ROS 会对癌组织造成显著的氧化损伤，进一步触发癌细胞的凋亡过程[113]。这种 ROS 水平的上升，可以由多种因素共同引发，包括细胞代谢活性的降低、线粒体电子转运蛋白的功能异常、炎症反应的加剧、缺氧环境的形成以及致癌信号的激活等。在体内，过氧化氢、超氧化物、单线态氧和羟基自由基等典型的 ROS 分子，不仅能够直接对细胞结构造成损伤，还可以作为触发机制，诱导药物在特定位置释放。与利用还原反应设计的纳米颗粒中常见的二硫键类似，酯键、硫缩酮键以及某些金属化合物也因其对 ROS 的响应性而被广泛应用于 ROS 反应型纳米颗粒的制备中[114]。这些纳米颗粒能够在 ROS 的作用下发生特定的化学变化，从而实现药物的精准释放和肿瘤治疗的效果提升。

Ni 等[115]采用“MOF-on-MOF”策略将阿霉素（Dox）封装在双层 NH_2-MIL-88 中，以制备核壳结构的载药 MOFs 进而实现 pH 和 GSH 双响应控制 Dox 释放。由于核壳结构的载药量达到 14.4%（质量分数），几乎是 MOFs 单独载药的两倍，同时核壳结构的控释性能也得到改善。

复杂的肿瘤微环境（TME）和非特异性药物靶向限制了光动力学治疗联合化疗的临床疗效。Wang 等据报道可通过减少肿瘤缺氧和细胞内谷胱甘肽（GSH）来调节 TME 金属有机框架（MOFs）辅助策略，并提供靶向递送和受控释放的化学药物。铂(Ⅳ)-二叠氮基络合物（Pt(Ⅳ)）装入基于羧酸铜(Ⅱ)的 MOF-199 中，并且聚集诱导发射光敏剂 TBD 与聚乙二醇偶联来封装 MOF-199。在光照射下，释放的 Pt(Ⅳ)产生 O_2 可缓解缺氧并在癌细胞内产生基于 Pt(Ⅱ)的化学药物[116]。

过氧化氢（H_2O_2）/谷胱甘肽（GSH）双敏感纳米平台有望减轻化疗药物的副作用并提高其对癌症的治疗效果。Miao 等开发了具有二硫键交联结构的聚甲基丙烯酸核（PMAA BACy）的新型多功能药物递送系统，此系统可用作按需药物释放的载体金属有机框架夹层和仿生聚多巴胺（PDA）涂层，双响应纳米平台不仅可以防止化疗药物的过早泄漏，而且对生物学相关的 GSH 和 H_2O_2 敏感，从而可以精确递送化疗药物[117]。

Du 等[118]使用谷胱甘肽敏感的金属有机框架制备纳米药物（BMS-SNAP-MOF），以封装免疫抑制酶吲哚胺 2,3-双加氧酶（IDO）抑制剂 BMS-986205 和一氧化氮（NO）供体。纳米药物通过 EPR 效应在肿瘤组织中积累并随后内化到肿瘤细胞中后，其中富集的 GSH 引发与 MOFs 的级联反应，MOFs 分解纳米药物以快速释放 IDO 抑制 BMS-986205 并产生丰富的 NO。

4.3.2　ATP 响应药物递送平台

三磷酸腺苷（adenosine triphosphate，ATP）是生物体内最直接的能量来源，是生化系统的核心。电化学生物传感作为一种主要的检测技术，可以对肿瘤细胞中的 ATP 进行定量检测[119]，从而有助于对肿瘤的研究治疗。目前，对肿瘤标志物的电化学生物传感检测是一项新的前沿研究热点[120]。而 ATP 与肿瘤疾病密切相关，因此，若能实现对 ATP 的高灵敏和高选择性检测，则会有利于肿瘤的早期检测和诊断。

据报道，MOFs 材料可被用作 ATP 响应型药物递送载体[121]。这是因为 ATP 的氨基、苯环以及咪唑环的 N 端含有孤对电子，这些孤对电子有可能与 MOFs 中的金属离子发生强烈的配位作用，并且这种配位作用具有可变的配位电位。另一方面，由于 ATP 中的 N 原子含有丰富的孤对电子，这使得 ATP-离子复合物与某些金属复合物之间展现出高亲和力[122]。基于这些特性，当 ATP 与 MOFs 的金属位点发生配位时，将会引发 MOFs 结构的裂解，并释放出所需的能量，从而实现药物的有效递送。

据研究表明，咪唑-2-甲醛和 Zn^{2+} 与蛋白质的自组装形成 ZIF-90/蛋白质纳米颗粒并有效地封装蛋白质。发现在 ATP 存在下，由于 ATP 与 ZIF-90 的 Zn^{2+}之间的竞争配位，ZIF-90/蛋白质纳米颗粒降解释放蛋白质[47]。

Willner 课题组研制了 MOFs 装载阿霉素（Dox）体系，释放药物分子阿霉素 [123]。也有研究采用一步合成法由氨基功能化配体和 Ce^{3+}构建三维镧系 Ce-MOFs[124]。通过 ATP 适配体-Ce-MOFs 的构建来检测 ATP。这是基于 MOFs 的电化学传感器检测 ATP 的创新。该方法已成功应用于肿瘤患者血清中 ATP 的检测[125]，在临床诊断中具有潜在的应用价值。

有研究表明，设计合成的 APZIF-90 材料展现出了 ATP 和 pH 双重反应性，这一特性使其在药物递送领域具有显著优势。具体而言，在生理 pH 水平下，APZIF-90 能够有效降低药物的过早释放，从而提高药物的稳定性和安全性。而在肿瘤细胞内，由于 ATP 和 pH 水平的变化，APZIF-90 能够更快速、更有效地触发药物释放，从而增强对肿瘤的治疗效果。进一步的体内毒性评价显示，ZIF-90 材料具有良好的生物相容性，这意味着它能够在生物体内安全地发挥作用，而不会引起明显的副作用[126]。同时，在给定剂量下，ZIF-90 对肝肾功能的影响也最小，这进一步证实了其在实际应用中的可行性和安全性。因此，APZIF-90 作为一种具有 ATP 和 pH 双重反应性的药物递送材料，有望在肿瘤治疗领域发挥重要作用，为改善患者的生活质量和延长生存期提供新的治疗策略。Yang 小组[121]报告了 ATP 响应型沸石咪唑酯框架（ZIF-90）已被确立为细胞溶质蛋白递送以及 CRISPR/Cas9 基因组编辑的通用平台。在这一应用中，咪唑-2-甲醛和 Zn^{2+}与蛋白质通过自组装的方式，形成了 ZIF-90/蛋白质纳米颗粒，从而有效地封装蛋白质。这种纳米颗粒不仅保留了蛋白质的活性，而且通过其独特的 ATP 响应性，能在特定条件下释放封装的蛋白质，实现了对蛋白质的精准递送和释放控制。因此，ZIF-90 为蛋白质药物和基因编辑工具的递送提供了一种高效且安全的方法，有望在未来为生物医药领域的发展开辟新的道路。

Yao 等[15]通过微波辅助的方法合成了水稳定的 MIL-101（Fe）-C_4H_4。当浸泡在阿霉素碱性水溶液中时，纳米 MOFs 充当纳米海绵实验负载能力高达 24.5%（质量分数），同时保持高达 98%的负载效率。此外，作为设计的聚（乙二醇–共–丙二醇）（F127）修饰的纳米粒子（F127-DOX-MIL）在谷胱甘肽（GSH）和 ATP 的刺激下可以分解。

4.3.3　pH 响应药物递送平台

肿瘤细胞可在肿瘤组织微环境中产生较高水平的乳酸加缺氧，肿瘤组织中的 pH 值普遍低于正常组织，这一现象在溶酶体和内体中也同样存在，其 pH 值通常在 4.5～6.5[114]。这种特定的组织特性为药物递送提供了新的策略。针对这一低 pH 环境，研究者们设计了一系列酸性敏感的药物递送体系，以实现 pH 响应性的药物释放。由于肿瘤组织的 pH 值约为 6.5，较正常组织（pH 值约为 7.4）更为酸性，这种药物递送体系能够在到达肿瘤区域时更为精准地释放药物。更进一步地，随着肿瘤的发展进入晚期，其内部的内体和溶酶体 pH 值会进一步降低[127]，通常处于 4.5～5.5 的范围。这种亚细胞水平的 pH 差异使得 pH 响应型药物递送体系能够更为精确地定位并释放药物于肿瘤细胞的特定部位，从而提高治疗效果并减少副作用。pH 响应药物递送平台也是目前应用最多的。要制备特定的 MOFs 材料应使其在生理条件下，这些材料能够保持稳定存在，但在酸性微环境中，其特性则发生显著变化。由于材料对酸敏感，会发生水解或化学基团的质子化过程，这导致材料结构坍塌或递送“开关”得以打开。一旦这些变化发生，封装的药物便能迅速释放至周围环境中。药物的快速释放使得细胞质中的药物浓度在短时间内急剧上升，远超过药物被细胞泵出的速率。这种高浓度的药物环境能够有效抑制细胞对药物的排出机制，从而显著缓解多药耐药性的问题[128]，提高治疗效果。因此，这种酸响应型药物递送系统为克服肿瘤治疗中的耐药性挑战提供了新的策略。

大量的研究尝试解释 MOFs 在酸和碱中的稳定性，基于高价金属离子和羧酸配体的 MOFs 在酸中比较稳定，对碱的抗性较低[129]。相反，基于软二价金属离子和氮基配体的 MOFs 在碱性溶液中更稳定，在酸中则更不稳定。沸石咪唑酯框架 ZIF-8 于 2006 年合成，是最典型的 pH 触发框架[130]。最初的合成负载 5-氟尿嘧啶（5-FU），并利用其对生理病理 pH 信号的反应[131]。装载 5-FU 后，在不同的 pH 条件下进行试验。在 pH 为 5.0 时，1 小时内的释放量接近磷酸盐缓冲盐水（PBS）的 2.7 倍。之后，已经有了许多关于基于 ZIF-8 的 pH 响应的 DDS 报告。

Liu 实验组以胶原蛋白酶（CLG）为核心，他们成功构建了包含 pH 敏感的苯甲亚胺配体（BI-linker）和 Mn^{2+}的片状 NCPs 纳米颗粒[132]。在弱酸性的肿瘤微环境（TME）中，这些 NCP@CLG-PEG 纳米颗粒能够响应环境的变化而发生解离，从而释放出 CLG。释放出的 CLG 能够特异性地降解 TME 中的胶原蛋白，有效降低肿瘤内部的间隙流体压力，进而改善血流灌注并缓解肿瘤乏氧状态。这一作用不仅有助于改善肿瘤组织的微环境，还能显著提高纳米制剂在肿瘤部位的富集效率。随着纳米颗粒在肿瘤区域的有效聚集，药物治疗的局部浓度得到显著提升，从而有效增强了治疗效果。因此，这种以 CLG 为中心的 pH 响应型 NCPs 纳米颗粒为肿瘤治疗提供了一种新的策略，有望在未来的临床应用中发挥重要作用。

有研究设计了基于多酶金属有机框架（MOFs）（胰岛素和葡萄糖氧化酶载钴的 ZIF-8，简称为 Ins/GOx@Co-ZIF-8）刺激响应微针（MNs），用于无痛葡萄糖介导的经皮给药[133]。酸敏感的 ZIF-8 被设计为过氧化氢酶模拟载体[134]，在胰岛素释放过程中分解过氧化氢。当 MNs 与高浓度血糖反应时，嵌入 MOFs 中的 GOx 将葡萄糖转化为葡萄糖酸。局部的酸性环境会导致 MOFs 的降解，并随后释放胰岛素。含有牙髓间充质干细胞膜的金属有机框架纳米颗粒 ZIF-8 可以携带 Dox 形成载药 MOFs，并被口腔鳞状细胞癌分泌的趋化因子化学吸引，从而抑制 OSCC 在体内外生长（图 4.4）[135]。而 MOF-Dox@DPSCM 在 pH 7.4 时的释放量为 43.2%±1.3%。在 pH 6.0 时，MOF-Dox@DPSCM 的 Dox 释放效率明显高于 pH 7.4，证明药物可以靶向释放[136]。

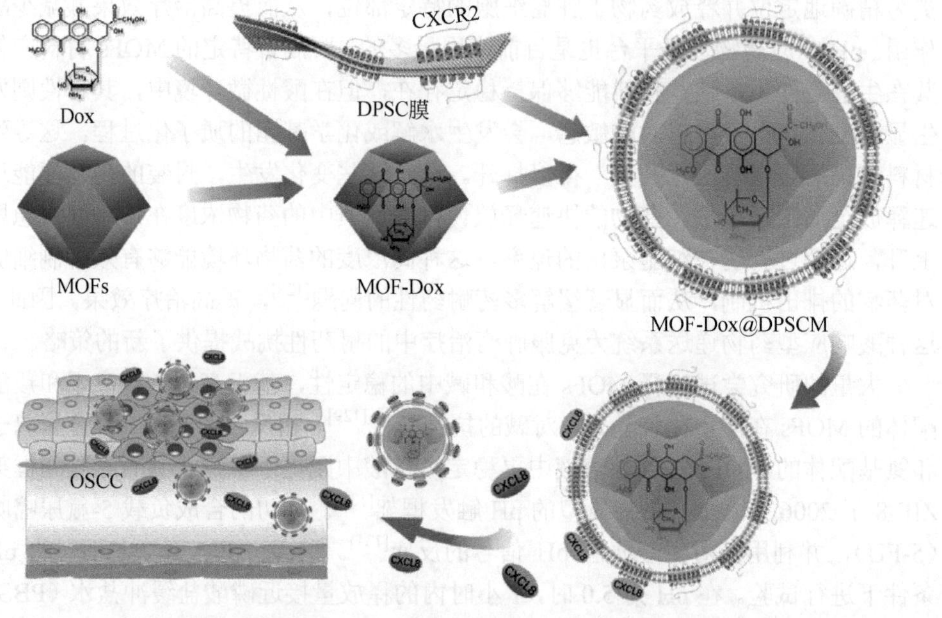

图 4.4　ZIF-8 携带 Dox 形成载药 MOFs，抑制 OSCC 在体内外生长

大多数药物载体的 MOFs 表现出正常的 pH 反应，由于框架的降解或/和主客体相互作用较弱，药物在酸性条件下更容易释放，因此不适合口服给药。有研究设计并合成了一种新的氮杂环有机配体 H_2QDDA，并利用其构建了第一个基于 Zr 的 MOFs（称为 ZJU-802）[137]。阴离子药物封装 ZJU-802 表现出一个罕见的反向 pH 响应，由于 N 的阴离子药物可以在正常情况下稳定释放，当 pH=2.0 由于增强框架和阴离子药物之间的相互作用保持稳定。这种反向药物释放非常适合口服给药[138]。此外，通过 ^1H NMR、MTT 检测和共聚焦显微镜成像，充分证明了 ZJU-802 的高负载能力（＞40%）、低细胞毒性和良好的生物相容性。

生物相容性 Ti 基金属有机框架 MIL-125 被探索作为一种潜在的 pH 控制药物递送载体，可在酸性条件下促进药物释放[139]。Reddy 等[140]用配体溶液对锌金属离子交联 PSH 进行后处理，生成了 nZIF-8@PAM/淀粉复合材料。这种新设计的 MOFs 水凝胶纳米结构被发现具有自黏性，还表现出更高的机械强度、黏弹性和 pH 响应行为。利用这些特性，它已被用作潜在光敏剂药物（玫瑰红）的缓释药物递送平台。

Jia 等[141]将 Dox 装载到 HMS 的内部空腔中，然后将沸石咪唑酯框架（ZIF-8）纳米颗粒涂在装有 Dox 的 HMS 的外表面上。获得的材料是胶囊（Dox/HMS @ ZIF），其中封装了 Dox。Dox/HMS @ ZIF 可以用作有效的 pH 响应药物递送系统。Dox 不会在生理条件下（pH 7.4）释放，而是在低 pH 值（4～6）下从 Dox/HMS @ ZIF 释放。Tang 等[142]以 ZIF-8 纳米颗粒为模板，其上的铁(III)-儿茶酚复合物薄膜涂层源自多巴胺改性藻酸盐（AlgDA）和铁(III)离子之间的配位。去除模板后，得到具有 pH 响应壁的纳米胶囊。阿霉素（Dox）是一种典型的抗癌药物，首先通过共沉淀固定在 ZIF-8 框架中，然后在去除 ZIF-8 后封装在纳米胶囊中。铁(III)-儿茶酚复合物的结构随 pH 值变化，从而赋予 Dox@Nanocapsules 体外定制释放行为。

关于 pH 响应型 MOFs 载药的研究较多，有可能向临床转化（表 4.1）。

表 4.1　一些 pH 响应型 MOFs 的总结

MOFs	二级构筑单元（SBU）	药品和装载能力 （质量分数/%）	药物释放 pH=5 pH=7.4	参考 文献
		Fluorouracil	75%	[143]
[Dy(HABA)(ABA)(DMA)$_4$]	Dy（III）	20.6	59%	[144]
NMOF	calcium（Ca^{2+}）	Zoledronate（Zol）	—	
ZIF-8	Zn^{2+}	Fluorouracil 45.4	45% 17%	[145]
ZIF-8@PAAS	Zn^{2+}	Doxorubicin 173～385	＞60% ＜15%	[146]
ZnO-DOX@ZIF-8	Zn^{2+}	Doxorubicin 11.2	80%（pH 5.5） 20%	[148]
CaZol	Ca^{2+}	Zoledronate 76	100% ＜5%	[144]
GMP/Eu	Eu^{3+}	OVA 55	60% ＜25%	[149]
PCN-221	Zr^{4+}	Methotrexate 40	40%（pH 2） 100%	[150]
PCN-221	Zr^{4+}	Methotrexate 0.40 g/g	35%（pH 2） 100%	[151]
ZJU-101	Zr^{4+}	Diclofenac sodium 54.6	—	[147]

4.3.4　缺氧响应药物递送平台

肿瘤细胞生长迅速，血液供应往往难以跟上其需求，导致缺氧成为肿瘤微环境中一个显著且重要的特征[152]。缺氧不仅在肿瘤侵袭、转移等恶性进程中起到关键作用，还对免疫系统产生抑制作用，加剧了肿瘤治疗的难度。近年来，随着对肿瘤微环境认识的深入，开发缺氧响应型纳米载体已成为研究热点，这种纳米载体能够在缺氧条件下发挥特定的药物递送或治疗作用[153]，为肿瘤治疗提供了新的策略和方向。

缺氧是肿瘤部位一个尤为关键的特征，它严重削弱了光敏剂的性能，进而影响了光动力学治疗（PDT）的效果。因此，研发一种能在缺氧条件下仍保持高光敏性的新型光敏剂，成为当前亟待解决的难题[154]。Zhang 课题组[155]最近报道了一种基于苯并卟啉的金属有机框架（TBP-MOF），其结构中包含 10 个 Zr 簇。相较于传统的基于卟啉的 MOFs，这种新型的 TBP-MOF 在光物理性能上有了显著的提升。它展现出了红移的吸收带和强烈的近红外发光特性，这使得它在生物成像领域具有潜在的应用价值。值得一提的是，TBP-MOF 中的π扩展苯并卟啉配体在促进超氧阴离子产生方面发挥了关键作用，从而显著增强了 PDT 的效果。研究结果显示，经过 PEG 修饰的纳米级 TBP-MOF（TBP-nMOF）能够在缺氧的肿瘤微环境中作为高效的 PDT 剂。这一成功合成的纳米平台不仅能够有效抑制原发肿瘤的生长，还能刺激抗肿瘤免疫反应，进而抑制转移性肿瘤的生长。这一发现为克服肿瘤治疗中的缺氧挑战提供了新的思路和方法，有望在未来的癌症治疗中发挥重要作用。

4.3.5　酶响应药物递送平台

肿瘤微环境中，由于肿瘤代谢和增殖的不平衡，某些酶的表达和活性会发生变化[156]。鉴于这些酶在多种病理和生理过程中的重要作用，研究者们开始致力于开发一种智能纳米载体，它能对这些酶产生响应并相应地发挥作用。酶作为响应触发物的优势在于其对底物具有高度特异性，这一特性使得酶响应纳米载体备受关注。这类纳米系统通过物理封装或共价连接的方式将药物载入纳米颗粒中，随后利用酶裂解等机制在特定部位释放药物。另外，某些系统还通过酶激活载体，使靶向配体暴露，从而更容易被靶向细胞吸收。在癌症治疗中，多种酶对药物释放起着至关重要的作用。其中，蛋白酶[157]、氧化还原酶[158]和磷脂酶[159]等酶类在肿瘤发展和治疗中扮演着重要角色。因此，设计和开发针对这些酶的响应型纳米载体，有望为癌症治疗提供新的策略和方法。

4.3.6　细胞线粒体响应药物递送平台

Zhang 课题组用肿瘤细胞膜修饰基于卟啉的 NMOFs，用于免疫逃逸和同源靶向[160]。肿瘤细胞表现出免疫耐受和肿瘤同源结合，它们与其特异性质膜蛋白密切相关。有趣的是，Zhang 等通过表面配位化学建立了 MOFs-DNA 功能化的通用方法，并进一步用 DNA 适配体（AS1411）修饰 PCN-224 nMOFs[161]，赋予 PCN-224 nMOFs 具有特异性靶向人乳腺癌细胞 MDA-MB-231 的特异性分子识别能力。

近期 Hu 课题组[162]报道了一种创新的混合药物递送系统（DDS）开发方法，该方法的关键在于将锌基金属有机框架（Zn-MOFs）与叶酸（FA）巧妙地结合。在此研究中，研究者们选取 Dox 作为模型抗癌药物，并将其高效地负载到 FA@Zn-MOF 纳米颗粒中。这种创新的混合 DDS 不仅提高了药物的稳定性，还增强了其在目标组织中的定向释放能力，为癌症治疗提供了新的可能。

Lin 课题组[163]注意到锌酞菁光敏剂（PSs）在长波长（650～750 nm）下展现出强大的吸收能力、高三重态量子产率以及良好的生物相容性，但它在体内易聚集并导致猝灭的问题一直限制着其应用。为了解决这一难题，他们设计了一种新型的纳米级金属有机层材料。这种材料能够有效地防止 MOFs 材料的激发态因聚集而发生自猝灭，从而高效地敏化单线态氧的形成，进而杀死肿瘤细胞。更为值得一提的是，这种材料具有足够的亲脂性和正电荷，这使得它能够在没有三苯基膦的辅助下就能精准地靶向线粒体。这一特性不仅提高了光敏剂的治疗效果，还降低了其在非靶标部位的副作用，为癌症治疗提供了一种新的高效、低毒的策略。

在近期的研究中 Alijani 等[164]将四氧化三铁核心 UiO-66-NH_2 MOF 壳纳米结构 Fe_3O_4@MOF 装载抗癌药物阿霉素（Dox）并将负载的纳米结构偶联到高荧光碳点（CDs）上，然后用核仁结合适配体 AS1411 覆盖用于治疗；为解决雌激素受体（ER）孕酮受体（PR）和人表皮生长因子受体不表达引起的乳腺癌，通过一个适配体偶联的 MOFs，将靶向药物作为一种三阴性的人乳腺癌细胞系传递到 MDA-MB-231。这种适配体功能化的碳点（CDs）封装的 Fe_3O_4@-MOF 核壳纳米结构可以用于改良肿瘤化疗。

参 考 文 献

[1] Karami A, Mohamed O, Ahmed A, et al. Recent advances in metal-organic frameworks as anticancer drug delivery systems: A review. Anti-Cancer Agents in Medicinal Chemistry, 2021, 21: 2487-2504.

[2] Wang Liying, Chen Yunching, Lin Hsinyao, et al. Near-Ir-absorbing gold nanoframes with enhanced physiological stability and improved biocompatibility for *in vivo* biomedical applications. ACS Applied Materials & Interfaces, 2017, 9: 3873-3884.

[3] Moharramnejad M, Ehsani A, Shahi M, et al. MOF as nanoscale drug delivery devices: Synthesis and recent progress in biomedical applications. Journal of Drug Delivery Science and Technology, 2023, 81: 104285.

[4] Zhu Wei, Xiang Guolei, Shang Jin, et al. Versatile surface functionalization of metal-organic

frameworks through direct metal coordination with a phenolic lipid enables diverse applications. Advanced Functional Materials, 2018, 28: 1705274.
[5] Morris R E, Brammer L. Coordination change, lability and hemilability in metal-organic frameworks. Chemical Society Reviews, 2017, 46: 5444-5462.
[6] Guo Chuanpan, Duan Fenghe, Zhang Shuai, et al. Heterostructured hybrids of metal-organic frameworks (MOFs) and covalent–organic frameworks (COFs). Journal of Materials Chemistry A, 2022, 10: 475-507.
[7] Yu Qi, Zou Jin, Yu Chenxiao, et al. Nitrogen doped porous biochar/B-Cd-MOFs heterostructures: Bi-functional material for highly sensitive electrochemical detection and removal of acetaminophen. Molecules, 2023, 28: 2437.
[8] Han Zhiwei, Fan Xinyang, Yu Shuyu, et al. Metal-organic frameworks (MOFs): A novel platform for laccase immobilization and application. Journal of Environmental Chemical Engineering, 2022, 10: 108795.
[9] Zhao Xudong, Zheng Meiqi, Gao Xinli, et al. The application of MOFs-based materials for antibacterials adsorption. Coordination Chemistry Reviews, 2021, 440: 213970.
[10] Yang Shengjiang, Zhao Daohui, Xu Zhiyong, et al. Molecular understanding of acetylcholinesterase adsorption on functionalized carbon nanotubes for enzymatic biosensors. Physical Chemistry Chemical Physics, 2022, 24: 2866-2878.
[11] Ma Wenjie, Jiang Qin, Yu Ping, et al. Zeolitic imidazolate framework-based electrochemical biosensor for *in vivo* electrochemical measurements. Analytical Chemistry, 2013, 85: 7550-7557.
[12] Feng Dawei, Liu Tianfu, Su Jie, et al. Stable metal-organic frameworks containing single-molecule traps for enzyme encapsulation. Nature Communications, 2015, 6: 5979.
[13] Peng Shuang, Bie Binglin, Sun Yangzesheng, et al. Metal-organic frameworks for precise inclusion of single-stranded DNA and transfection in immune cells. Nature Communications, 2018, 9: 1293.
[14] Yin Xiao, Yang Bin, Chen Beibei, et al. Multifunctional gold nanocluster decorated metal-organic framework for real-time monitoring of targeted drug delivery and quantitative evaluation of cellular therapeutic response. Analytical Chemistry, 2019, 91: 10596-10603.
[15] Yao Lijia, Tang Ying, Cao Wenqian, et al. Highly efficient encapsulation of doxorubicin hydrochloride in metal-organic frameworks for synergistic chemotherapy and chemodynamic therapy. ACS Biomaterials Science & Engineering, 2021, 7: 4999-5006.
[16] Liu Jintong, Zhang Lei, Lei Jianping, et al. Multifunctional metal-organic framework nanoprobe for cathepsin B-activated cancer cell imaging and chemo-photodynamic therapy. ACS Applied Materials & Interfaces, 2017, 9: 2150-2158.

[17] Soomro N A, Wu Qiao, Amur S A, et al. Natural drug physcion encapsulated zeolitic imidazolate framework, and their application as antimicrobial agent. Colloids and Surfaces B: Biointerfaces, 2019, 182: 110364.

[18] Ding Meili, Cai Xuechao, Jiang Hailong. Improving MOF stability: Approaches and applications. Chemical Science, 2019, 10: 10209-10230.

[19] Zou Zhen, Li Siqi, He Dinggeng, et al. A versatile stimulus-responsive metal-organic framework for size/morphology tunable hollow mesoporous silica and pH-triggered drug delivery. Journal of Materials Chemistry B, 2017, 5: 2126-2132.

[20] Zhuang Jia, Kuo Chunhong, Chou Lienyang, et al. Optimized metal-organic-framework nanospheres for drug delivery: Evaluation of small-molecule encapsulation. ACS Nano, 2014, 8: 2812-2819.

[21] Ding He, Song Yang, Huang Xiaowan, et al. Mtorc1-dependent TFEB nucleus translocation and pro-survival autophagy induced by zeolitic imidazolate framework-8. Biomaterials Science, 2020, 8: 4358-4369.

[22] Alves R C, Schulte Z M, Luiz M T, et al. Breast cancer targeting of a drug delivery system through postsynthetic modification of curcumin@N_3-Bio-MOF-100 via click chemistry. Inorganic Chemistry, 2021, 60: 11739-11744.

[23] Han Bo, Chakraborty A. Highly efficient adsorption desalination employing protonated-amino-functionalized MOFs. Desalination, 2022, 541: 116045.

[24] Huang Yu, Li Yaru, Zhang Dongsheng, et al. Light-switchable N-alkylation using amine-functionalized MOF. Applied Catalysis B: Environment and Energy, 2024, 350: 123924.

[25] Cai Zhongzheng, Tao Wenjie, Moore C E, et al. Direct no reduction by a biomimetic iron（Ⅱ）pyrazolate MOF. Angewandte Chemie International Edition, 2021, 60: 21221-21225.

[26] Wei Zihao, Xu Wenquan, Peng Panpan, et al. Covalent synthesis of Ti-MOF for enhanced photocatalytic CO_2 reduction. Molecular Catalysis, 2024, 558: 114042.

[27] Pang Jiandong, Yuan Shuai, Du Dongying, et al. Flexible zirconium MOFs as bromine-nanocontainers for bromination reactions under ambient conditions. Angewandte Chemie International Edition, 2017, 56: 14622-14626.

[28] Anbardan S Z, Mokhtari J, Yari A, et al. Direct synthesis of amides and imines by dehydrogenative homo or cross-coupling of amines and alcohols catalyzed by Cu-MOF. RSC Advances, 2021, 11: 20788-20793.

[29] Gong Linshan, Zhu Juncheng, Yang Yuxin, et al. Effect of polyethylene glycol on polysaccharides: From molecular modification, composite matrixes, synergetic properties to embeddable application in food fields. Carbohydrate Polymers, 2024, 327: 121647.

[30] Yan Hongde, Guo Binghan, Wang Zhao, et al. Surfactant-modified aspergillus oryzae lipase as

a highly active and enantioselective catalyst for the kinetic resolution of (Rs)-1-phenylethanol. 3 Biotech, 2019, 9: 265.

[31] Samui A, Chowdhuri A R, Mahto T K, et al. Fabrication of a magnetic nanoparticle embedded NH_2-Mil-88b MOF hybrid for highly efficient covalent immobilization of lipase. RSC Advances, 2016, 6: 66385-66393.

[32] Lin Chunhua, Du Yue, Wang Shiqi, et al. Glucose oxidase@Cu-hemin metal-organic framework for colorimetric analysis of glucose. Materials Science and Engineering: C, 2021, 118: 111511.

[33] Chambre L, Maouati H, Oz Y, et al. Thiol-reactive clickable cryogels: Importance of macroporosity and linkers on biomolecular immobilization. Bioconjugate Chemistry, 2020, 31: 2116-2124.

[34] Morris W, Briley W E, Auyeung E, et al. Nucleic acid-metal organic framework (MOF) nanoparticle conjugates. Journal of the American Chemical Society, 2014, 136: 7261-7264.

[35] Wang Shunzhi, McGuirk C M, Ross M B, et al. General and direct method for preparing oligonucleotide-functionalized metal-organic framework nanoparticles. Journal of the American Chemical Society, 2017, 139: 9827-9830.

[36] Wang Yutong, Feng Liang, Zhang Kai, et al. Uncovering structural opportunities for zirconium metal-organic frameworks via linker desymmetrization. Advanced Science, 2019, 6: 1901855.

[37] Bhat Z U H, Hanif S, Rafi Z, et al. New mixed-ligand Zn(Ⅱ)-based MOF as a nanocarrier platform for improved antibacterial activity of clinically approved drug levofloxacin. New Journal of Chemistry, 2023, 47: 7416-7424.

[38] He Zehui, Zhang Yongtai, Liu Zhenda, et al. Synergistic treatment of androgenetic alopecia with follicular Co-delivery of minoxidil and cedrol in metal-organic frameworks stabilized by covalently cross-linked cyclodextrins. International Journal of Pharmaceutics, 2024, 654: 123948.

[39] Subramaniyam V, Ravi P V, Pichumani M. Structure Co-ordination of solitary amino acids as ligands in metal-organic frameworks (MOFs): A comprehensive review. Journal of Molecular Structure, 2022, 1251: 131931.

[40] Zhao Zhanfeng, Yang Dong, Ren Hanjie, et al. Nitrogenase-inspired mixed-valence MIL-53(Fe II/Fe III) for photocatalytic nitrogen fixation. Chemical Engineering Journal, 2020, 400: 125929.

[41] Sun Baoting, Bilal M, Jia Shiru, et al. Design and bio-applications of biological metal-organic frameworks. Korean Journal of Chemical Engineering, 2019, 36: 1949-1964.

[42] Shen Xiang, Yan Bing. Polymer hybrid thin films based on rare earth ion-functionalized MOF: Photoluminescence tuning and sensing as a thermometer. Dalton Transactions, 2015, 44:

1875-1881.
[43] Salama E, Ghanim M, Hassan H S, et al. Novel aspartic-based Bio-MOF adsorbent for effective anionic dye decontamination from polluted water. RSC Advances, 2022, 12: 18363-18372.
[44] Pander M, Żelichowska A, Bury W. Probing mesoporous Zr-MOF as drug delivery system for carboxylate functionalized molecules. Polyhedron, 2018, 156: 131-137.
[45] Yang Guangbao, Liu Jingjing, Wu Yifan, et al. Near-infrared-light responsive nanoscale drug delivery systems for cancer treatment. Coordination Chemistry Reviews, 2016, 320-321: 100-117.
[46] Xu Shibo, Lin Yiyi, Zhao Xingjun, et al. Nir triggered photocatalytic and photothermal bifunctional MOF nanozyme using for improving osteoarthritis microenvironment by repairing injured chondrocytes of mitochondria. Chemical Engineering Journal, 2023, 468: 143826.
[47] Sheng Chuangui, Yu Fangzhi, Feng Youming, et al. Near-infrared light triggered degradation of metal-organic frameworks for spatiotemporally-controlled protein release. Nano Today, 2023, 49: 101821.
[48] Yu Hongbo, Xu Xiaomu, Xie Zheng, et al. High-efficiency near-infrared light responsive antibacterial system for synergistic ablation of bacteria and biofilm. ACS Applied Materials & Interfaces, 2022, 14: 36947-36956.
[49] Ruan Yushan, Chen Lineng, Cui Lianmeng, et al. Ppy-modified prussian blue cathode materials for low-cost and cycling-stable aqueous zinc-based hybrid battery. Coatings, 2022, 12: 779.
[50] Zhang Luyun, Gao Yang, Sun Sijia, et al. pH-responsive metal-organic framework encapsulated gold nanoclusters with modulated release to enhance photodynamic therapy/chemotherapy in breast cancer. Journal of Materials Chemistry B, 2020, 8: 1739-1747.
[51] Li Cheng, Zheng Xianchuang, Chen Weizhi, et al. Tumor microenvironment-regulated and reported nanoparticles for overcoming the self-confinement of multiple photodynamic therapy. Nano Letters, 2020, 20: 6526-6534.
[52] Zhao Jintao, Wang Zihua, Zhong Miao, et al. Integration of a diselenide unit generates fluorogenic camptothecin prodrugs with improved cytotoxicity to cancer cells. Journal of Medicinal Chemistry, 2021, 64: 17979-17991.
[53] Yang Wei, Leng Tianchi, Miao Weicheng, et al. Photo-switchable peroxidase/catalase-like activity of carbon quantum dots. Angewandte Chemie International Edition, 2024, 63: e202403581.
[54] Yang Zhiyao, Liu Zejiang, Yuan Lihua. Recent advances of photoresponsive supramolecular switches. Asian Journal of Organic Chemistry, 2021, 10: 74-90.
[55] Kaur R, Kim K H, Deep A. A convenient electrolytic assembly of graphene-MOF composite

thin film and its photoanodic application. Applied Surface Science, 2017, 396: 1303-1309.

[56] Musarurwa H, Tavengwa N T. Smart metal-organic framework (MOF) composites and their applications in environmental remediation. Materials Today Communications, 2022, 33: 104823.

[57] Mahato M, Hwang W J, Tabassian R, et al. A dual-responsive magnetoactive and electro–ionic soft actuator derived from a nickel-based metal-organic framework. Advanced Materials, 2022, 34: 2203613.

[58] Akbar M U, Badar M, Zaheer M. Programmable drug release from a dual-stimuli responsive magnetic metal-organic framework. ACS Omega, 2022, 7: 32588-32598.

[59] Gao Xuechuan, Ji Guanfeng, Cui Ruixue, et al. Controlled synthesis of MOFs@MOFs core–shell structure for photodynamic therapy and magnetic resonance imaging. Materials Letters, 2019, 237: 197-199.

[60] Meng Ling, Ren Na, Dong Mengwei, et al. Metal-organic frameworks for nerve repair and neural stem cell therapy. Advanced Functional Materials, 2024, 34: 2309974.

[61] Din M I, Sadaf S, Hussain Z, et al. Assembly of superparamagnetic iron oxide nanoparticles（Fe_3O_4-Nps） for catalytic pyrolysis of corn cob biomass. Energy Sources, Part A: Recovery, Utilization, and Environmental Effects, 2020, 22: 1-9.

[62] Sun Z K, Zhou X R, Luo W, et al. Interfacial engineering of magnetic particles with porous shells: Towards magnetic core - porous shell microparticles. Nano Today, 2016, 11: 464-482.

[63] Shylesh S, Wang Lei, Thiel W R. Palladium(II)-phosphine complexes supported on magnetic nanoparticles: Filtration-free, recyclable catalysts for suzuki–miyaura cross-coupling reactions. Advanced Synthesis & Catalysis, 2010, 352: 425-432.

[64] Wu Dongfang, Takahashi K, Fujibayashi M, et al. Fluoride-bridged dinuclear dysprosium complex showing single-molecule magnetic behavior: Supramolecular approach to isolate magnetic molecules. RSC Advances, 2022, 12: 21280-21286.

[65] Zhao Kai, Gong Peiwei, Huang Jie, et al. Fluorescence turn-off magnetic cof composite as a novel nanocarrier for drug loading and targeted delivery. Microporous and Mesoporous Materials, 2021, 311: 110713.

[66] Sim T, Choi B, Kwon S W, et al. Magneto-activation and magnetic resonance imaging of natural killer cells labeled with magnetic nanocomplexes for the treatment of solid tumors. ACS Nano, 2021, 15: 12780-12793.

[67] Li Sheng, Bi Ke, Xiao Ling, et al. Facile preparation of magnetic metal organic frameworks core-shell nanoparticles for stimuli-responsive drug carrier. Nanotechnology, 2017, 28: 495601.

[68] Bhattacharjee A, Gumma S, Purkait M K. Fe_3O_4 promoted metal organic framework MIL-100(Fe) for the controlled release of doxorubicin hydrochloride. Microporous and

Mesoporous Materials, 2018, 259: 203-210.

[69] Yang Yongmei, Xia Feng, Yang Ying, et al. Litchi-like Fe_3O_4@Fe-MOF capped with Hap gatekeepers for pH-triggered drug release and anticancer effect. Journal of Materials Chemistry B, 2017, 5: 8600-8606.

[70] James S L. MOFs prepared by solvent-free grinding reactions. Acta Crystallographica A-Foundation and Advances, 2009, 65: S100.

[71] Ke Fei, Jiang Jing, Li Yizhi, et al. Highly selective removal of Hg^{2+} and Pb^{2+} by thiol-functionalized Fe_3O_4@metal-organic framework core-shell magnetic microspheres. Applied Surface Science, 2017, 413: 266-274.

[72] Chung C W, Liao Bowen, Huang Shuwei, et al. Magnetic responsive release of nitric oxide from an MOF-derived Fe_3O_4@Plga microsphere for the treatment of bacteria-infected cutaneous wound. ACS Applied Materials & Interfaces, 2022, 14: 6343-6357.

[73] Wu Wei, Liu Jianxi, Gong Peiwei, et al. Construction of core-shell NanoMOFs@microgel for aqueous lubrication and thermal-responsive drug release. Small, 2022, 18: 2202510.

[74] Zhang Xiaodong, Xiang Shang, Du Quanxin, et al. Effect of calcination temperature on the structure and performance of rod-like mnceox derived from MOFs catalysts. Molecular Catalysis, 2022, 522: 112226.

[75] Yurtseven H, Dogan E K. Magnetic, thermal and ferroelectric properties of MOFs (Mhym, M = Fe, Mn) close to phase transitions. Journal of Magnetism and Magnetic Materials, 2021, 540: 168489.

[76] Nakayama M, Okano T. Intelligent thermoresponsive polymeric micelles for targeted drug delivery. Journal of Drug Delivery Science and Technology, 2006, 16: 35-44.

[77] Ernst M, Poręba T, Gnägi L, et al. Locating guest molecules inside metal-organic framework pores with a multilevel computational approach. The Journal of Physical Chemistry C, 2023, 127: 523-531.

[78] Golden E, Yu Lijuan, Meilleur F, et al. An extended N-H bond, driven by a conserved second-order interaction, orients the flavin N5 orbital in cholesterol oxidase. Scientific Reports, 2017, 7: 40517.

[79] Cherroret N, Delande D, Tiggelen B A V. Induced dipole-dipole interactions in light diffusion from point dipoles. Physical Review A, 2016, 94: 012702.

[80] Sheridan R P, Lee R H, Peters N, et al. Hydrogen-bond cooperativity in protein secondary structure. Biopolymers, 1979, 18: 2451-2458.

[81] Li Mengyang, Zhou Yuqi, Wei Bing, et al. Insight into the interaction of host-guest structures for pyrrole-based metal compounds and C_{70}. The Journal of Chemical Physics, 2024, 160: 124307.

[82] Jiang Ke, Zhang Ling, Hu Quan, et al. Thermal stimuli-triggered drug release from a biocompatible porous metal-organic framework. Chemistry - A European Journal, 2017, 23: 10215-10221.

[83] Wang Jun, Huang Na, Peng Qi, et al. Temperature/pH dual-responsive and luminescent drug carrier based on pnipam-maa/lanthanide-polyoxometalates for controlled drug delivery and imaging in hela cells. Materials Chemistry and Physics, 2020, 239: 121994.

[84] Wang Jun, Chen Si, Cheng Xiaoyan, et al. Temperature/pH dual-responsive poly（N-isopropylacrylamide）/chitosan-coated luminescent composite nanospheres: Fabrication and controllable luminescence. Optical Materials, 2018, 86: 56-61.

[85] Tamura A, Osawa M, Yui N. Supermolecule-drug conjugates based on acid-degradable polyrotaxanes for pH-dependent intracellular release of doxorubicin. Molecules, 2023, 28: 2517.

[86] Nannuri S H, Pandey A, Kulkarni S, et al. A new paradigm in biosensing: MOF-carbon dot conjugates. Materials Today Communications, 2023, 35: 106340.

[87] Nagata S, Kokado K, Sada K. Metal-organic framework tethering pH and thermo-responsive polymer for on-off controlled release of guest molecules. Crystengcomm, 2020, 22: 1106-1111.

[88] Zhang Keju, Zhou Dong, Wang Zhiguo, et al. Hybrid mesoporous silica nanospheres modified by poly （Nipam-Co-Aa） for drug delivery. Nanotechnology, 2019, 30: 355604.

[89] Abdi J, Sisi A J, Hadipoor M, et al. State of the art on the ultrasonic-assisted removal of environmental pollutants using metal-organic frameworks. Journal of Hazardous Materials, 2022, 424: 127558.

[90] Yin Zhaoyang, Le Qichi, Zhou Weiyang, et al. Study on structural variation of Sn-20% Pb alloy melt subjected to ultrasonic vibration: An electrical characterization. Metals and Materials International, 2024, 29: 1-12.

[91] Wu Hao, Zhai Qingxi, Ding Fan, et al. Amorphous fenicu-MOFs as highly efficient electrocatalysts for the oxygen evolution reaction in an alkaline medium. Dalton Transactions, 2022, 51: 14306-14316.

[92] Zhang Lei, Zhang Jiahui, Sun Hui, et al. Enhanced photocatalytic performance of carbon fiber paper supported TiO_2 under the ultrasonic synergy effect. RSC Advances, 2022, 12: 22922-22930.

[93] Liu Xia, Zheng Tengfei, Wang Chaohui. Three-dimensional modeling and experimentation of microfluidic devices driven by surface acoustic wave. Ultrasonics, 2023, 129: 106914.

[94] Ahmed A, Karami A, Sabouni R, et al. pH and ultrasound dual-responsive drug delivery system based on peg–folate-functionalized iron-based metal-organic framework for targeted doxorubicin delivery. Colloids and Surfaces A: Physicochemical and Engineering Aspects,

2021, 626: 127062.

[95] Baig M M F A, Lai Wingfu, Ahsan A, et al. Synthesis of ligand functionalized Erbb-3 targeted novel DNA nano-threads loaded with the low dose of doxorubicin for efficient in vitro evaluation of the resistant anti-cancer activity. Pharmaceutical Research, 2020, 37: 75.

[96] Sun Ao. Applications of MOFs in drug delivery. Highlights in Science, Engineering and Technology, 2023, 58: 351-357.

[97] Liu Jianbo, Shang Yonghui, Zhu Qiuyan, et al. A voltammetric immunoassay for the carcinoembryonic antigen using silver(I)-terephthalate metal-organic frameworks containing gold nanoparticles as a signal probe. Microchimica Acta, 2019, 186: 509.

[98] Wang Huaisong, Wang Yihui, Ding Ya. Development of biological metal-organic frameworks designed for biomedical applications: From bio-sensing/bio-imaging to disease treatment. Nanoscale Advances, 2020, 2: 3788-3797.

[99] Mirhadi E, Mashreghi M, Maleki M F, et al. Redox-sensitive nanoscale drug delivery systems for cancer treatment. International Journal of Pharmaceutics, 2020, 589: 119882.

[100] Alkhatib Y, Blume G, Thamm J, et al. Overcoming the hydrophilicity of bacterial nanocellulose: Incorporation of the lipophilic coenzyme Q10 using lipid nanocarriers for dermal applications. European Journal of Pharmaceutics and Biopharmaceutics, 2021, 158: 106-112.

[101] Yang Yunlong, Li Yan, Lin Qiuning, et al. *In situ* phototriggered disulfide-cross-link nanoparticles for drug delivery. ACS Macro Letters, 2016, 5: 301-305.

[102] Shi Zhenfeng, Liu Jifang, Tian Lei, et al. Insights into stimuli-responsive diselenide bonds utilized in drug delivery systems for cancer therapy. Biomedicine & Pharmacotherapy, 2022, 155: 113707.

[103] Chang M J, Kim K, Kang C, et al. Enhanced aggregability of aie-based probe through H_2S-selective triggered dimerization and its applications to biological systems. ACS Omega, 2019, 4: 7176-7181.

[104] Park J, Choi Y, Chang H, et al. Alliance with EPR effect: Combined strategies to improve the EPR effect in the tumor microenvironment. Theranostics, 2019, 9: 8073-8090.

[105] Khalife K H, Lupidi G. Reduction of hypervalent states of myoglobin and hemoglobin to their ferrous forms by thymoquinone: The role of Gsh, Nadh and Nadph. Biochimica et Biophysica Acta （BBA） - General Subjects, 2008, 1780: 627-637.

[106] Lv Ruihong, Li Xinyi, Song Shuhui, et al. Fabrication and characterization of dual-responsive nanocarriers for effective drug delivery and synergistic chem-photothermal effects. Colloids and Surfaces A: Physicochemical and Engineering Aspects, 2022, 655: 130256.

[107] Yi Qiangying, Ma Jin, Kang Ke, et al. Dual cellular stimuli-responsive hydrogel nanocapsules

for delivery of anticancer drugs. Journal of Materials Chemistry B, 2016, 4: 4922-4933.

[108] Wu Qiong, Du Qijun, Sun Xiaohan, et al. MnMOF-based microwave-glutathione dual-responsive nano-missile for enhanced microwave thermo-dynamic chemotherapy of drug-resistant tumors. Chemical Engineering Journal, 2022, 439: 135582.

[109] Zhao Jiayue, Yang Yu, Han Xiao, et al. Redox-sensitive nanoscale coordination polymers for drug delivery and cancer theranostics. ACS Applied Materials & Interfaces, 2017, 9: 23555-23563.

[110] Lin Xiahui, Zhu Rong, Hong Zhongzhu, et al. Gsh-responsive radiosensitizers with deep penetration ability for multimodal imaging-guided synergistic radio-chemodynamic cancer therapy. Advanced Functional Materials, 2021, 31: 2101278.

[111] Cao Shuhua, Li Xuezhao, Gao Yong, et al. A simultaneously Gsh-depleted bimetallic Cu(II) complex for enhanced chemodynamic cancer therapy. Dalton Transactions, 2020, 49: 11851-11858.

[112] Wan Shuangshuang, Cheng Qian, Zeng Xuan, et al. A Mn(III)-sealed metal-organic framework nanosystem for redox-unlocked tumor theranostics. ACS Nano, 2019, 13: 6561-6571.

[113] Sun Songjia, Wang Dianwei, Yin Renyong, et al. A two-in-one nanoprodrug for photoacoustic imaging-guided enhanced sonodynamic therapy. Small, 2022, 18: 2202558.

[114] Kamil A, Khan M A, Asim M, et al. Detection of ROS and translocation of erp-57 in apoptotic induced human neuroblastoma (Sh-Sy5y) cells. Biocell, 2019, 43: 167.

[115] Ni Weishu, Zhang Ling, Zhang Hengrui, et al. Hierarchical MOF-on-MOF architecture for Ph/Gsh-controlled drug delivery and Fe-based chemodynamic therapy. Inorganic Chemistry, 2022, 61: 3281-3287.

[116] Wang Yuanbo, Wu Wenbo, Mao Duo, et al. Metal-organic framework assisted and tumor microenvironment modulated synergistic image-guided photo-chemo therapy. Advanced Functional Materials, 2020, 30: 2002431.

[117] Miao Yalei, Zhao Xubo, Qiu Yudian, et al. Metal-organic framework-assisted nanoplatform with hydrogen peroxide/glutathione dual-sensitive on-demand drug release for targeting tumors and their microenvironment. ACS Applied Bio Materials, 2019, 2: 895-905.

[118] Du Lihua, He Haozhe, Xiao Zecong, et al. Gsh-responsive metal-organic framework for intratumoral release of no and Ido inhibitor to enhance antitumor immunotherapy. Small, 2022, 18: 2107732.

[119] Khojastehnezhad A, Taghavi F, Yaghoobi E, et al. Recent achievements and advances in optical and electrochemical aptasensing detection of ATP based on quantum dots. Talanta, 2021, 235: 122753.

[120] Shi Jiaju, Chen Zichao, Zhao Chunqin, et al. Photoelectrochemical biosensing platforms for tumor marker detection. Coordination Chemistry Reviews, 2022, 469: 214675.

[121] Yang Xiaoti, Tang Qiao, Jiang Ying, et al. Nanoscale ATP-responsive zeolitic imidazole framework-90 as a general platform for cytosolic protein delivery and genome editing. Journal of the American Chemical Society, 2019, 141: 3782-3786.

[122] Molla G S, Himmelspach A, Wohlgemuth R, et al. Mechanistic and kinetics elucidation of Mg^{2+}/ATP molar ratio effect on glycerol kinase. Molecular Catalysis, 2018, 445: 36-42.

[123] Chen Weihai, Yu Xu, Liao Weiching, et al. ATP-responsive aptamer-based metal-organic framework nanoparticles (NMOFs) for the controlled release of loads and drugs. Advanced Functional Materials, 2017, 27: 1702102.

[124] Wang Shanshan, Xu Jin, Wang Juan, et al. Luminescence of samarium(III) bis-dithiocarbamate frameworks: Codoped lanthanide emitters that cover visible and near-infrared domains. Journal of Materials Chemistry C, 2017, 5: 6620-6628.

[125] Panikar S S, Banu N, Haramati J, et al. Anti-fouling sers-based immunosensor for point-of-care detection of the B7-H6 tumor biomarker in cervical cancer patient serum. Analytica Chimica Acta, 2020, 1138: 110-122.

[126] Jiang Zhenqi, Wang Yinjie, Sun Li, et al. Dual ATP and pH responsive ZIF-90 nanosystem with favorable biocompatibility and facile post-modification improves therapeutic outcomes of triple negative breast cancer *in vivo*. Biomaterials, 2019, 197: 41-50.

[127] Mao Jie, Li Yang, Wu Tong, et al. A simple dual-pH responsive prodrug-based polymeric micelles for drug delivery. ACS Applied Materials & Interfaces, 2016, 8: 17109-17117.

[128] Chen Zhijie, Kirlikovali K O, Shi Le, et al. Rational design of stable functional metal-organic frameworks. Materials Horizons, 2023, 10: 3257-3268.

[129] Wang Liangjie, Li Juan, Cheng Luyao, et al. Application of hard and soft acid base theory to uncover the destructiveness of lewis bases to UiO-66 type metal organic frameworks in aqueous solutions. Journal of Materials Chemistry A, 2021, 9: 14868.

[130] Nabipour H, Sadr M H, Bardajee G R. Synthesis and characterization of nanoscale zeolitic imidazolate frameworks with ciprofloxacin and their applications as antimicrobial agents. New Journal of Chemistry, 2017, 41: 7364-7370.

[131] Ajmi M F A, Hussain A, Ahmed F, et al. Novel synthesis of ZnO nanoparticles and their enhanced anticancer activity: role of ZnO as a drug carrier. Ceramics International, 2016, 42: 4462-4469.

[132] Liu Jingjing, Tian Longlong, Zhang Rui, et al. Collagenase-encapsulated pH-responsive nanoscale coordination polymers for tumor microenvironment modulation and enhanced photodynamic nanomedicine. ACS Applied Materials & Interfaces, 2018, 10: 43493-43502.

[133] Yang Xiaoxi, Feng Pengfei, Cao Jing, et al. Composition-engineered metal-organic framework-based microneedles for glucose-mediated transdermal insulin delivery. ACS Applied Materials & Interfaces, 2020, 12: 13613-13621.

[134] Guo Feng, Xu Zhonghao, Zhang Wendong, et al. Facile synthesis of catalase@ZIF-8 composite by biomimetic mineralization for efficient biocatalysis. Bioprocess and Biosystems Engineering, 2021 44: 1309-1319.

[135] Zhou Dabo, Chen Yixin, Bu Wenhuan, et al. Modification of metal-organic framework nanoparticles using dental pulp mesenchymal stem cell membranes to target oral squamous cell carcinoma. Journal of Colloid and Interface Science, 2021, 601: 650-660.

[136] Lázaro I A, Wells C J R, Ross S F, et al. Multivariate modulation of the Zr MOF Uio-66 for defect-controlled combination anticancer drug delivery. Angewandte Chemie International Edition, 2020, 59: 5211-5217.

[137] Jiang Ke, Zhang Ling, Hu Quan, et al. A zirconium-based metal-organic framework with encapsulated anionic drug for uncommonly controlled oral drug delivery. Microporous and Mesoporous Materials, 2019, 275: 229-234.

[138] Zhao Liwei, Gong Ming, Yang Jian, et al. Switchable ionic transportation in the nanochannels of the MOFs triggered by light and pH. Langmuir, 2021, 37: 13952-13960.

[139] Jiang Ke, Zhang Ling, Hu Quan, et al. A biocompatible Ti-based metal-organic framework for pH responsive drug delivery. Materials Letters, 2018, 225 142-144.

[140] Reddy Y N, De A, Paul S, et al. In situ nanoarchitectonics of a MOF hydrogel: A self-adhesive and Ph-responsive smart platform for phototherapeutic delivery. Biomacromolecules, 2023, 24: 1717-1730.

[141] Jia Xiaomin, Yang Zhiyuan, Wang Yujun, et al. Hollow mesoporous silica@metal-organic framework and applications for pH-responsive drug delivery. ChemMedChem, 2018, 13: 400-405.

[142] Tang Lei, Shi Jiafu, Wang Xiaoli, et al. Coordination polymer nanocapsules prepared using metal-organic framework templates for pH-responsive drug delivery. Nanotechnology, 2017, 28: 275601.

[143] Guo Yong, Yan Bing, Cheng Yu, et al. A new Dy(III)-based metal-organic framework with polar pores for pH-controlled anticancer drug delivery and inhibiting human osteosarcoma cells. Journal of Coordination Chemistry, 2019, 72: 262-271.

[144] Au K M, Satterlee A, Min Yuanzeng, et al. Folate-targeted Ph-responsive calcium zoledronate nanoscale metal-organic frameworks: turning a bone antiresorptive agent into an anticancer therapeutic. Biomaterials, 2016, 82: 178-193.

[145] Sun Chunyi, Qin Chao, Wang Xinlong, et al. Zeolitic imidazolate framework-8 as efficient

pH-sensitive drug delivery vehicle. Dalton Transactions, 2012, 41: 6906-6909.

[146] Yan Li, Chen Xianfeng, Wang Zhigang, et al. Size controllable and surface tunable zeolitic imidazolate framework-8-poly (acrylic acid sodium aalt) nanocomposites for pH responsive drug release and enhanced in Vivo cancer treatment. ACS Applied Materials & Interfaces, 2017, 9: 32990-33000.

[147] Yang Yanyu, Hu Quan, Zhang Qi, et al. A large capacity cationic metal-organic framework nanocarrier for physiological pH responsive drug delivery. Molecular Pharmaceutics, 2016, 13: 2782-2786.

[148] Zheng Cunchuan, Wang Yang, Soo Z F P, et al. ZnO-Dox@ZIF-8 core–shell nanoparticles for pH-responsive drug delivery. ACS Biomaterials Science & Engineering, 2017, 3: 2223-2229.

[149] Duan Fei, Feng Xiaochen, Yang Xinjian, et al. A simple and powerful Co-delivery system based on pH-responsive metal-organic frameworks for enhanced cancer immunotherapy. Biomaterials, 2017, 122: 23-33.

[150] Lin Wenxin, Hu Quan, Yu Jiancan, et al. Low cytotoxic metal-organic frameworks as temperature-responsive drug carriers. ChemPlusChem, 2016, 81: 668.

[151] Lin Wenxin, Hu Quan, Jiang Ke, et al. A porphyrin-based metal-organic framework as a pH-responsive drug carrier. Journal of Solid State Chemistry, 2016, 237: 307-312.

[152] Chang Cheng, Tang Xin, Daniel M, et al. Lrp-1 receptor combines Egfr signalling and Ehsp90α autocrine to support constitutive breast cancer cell motility in absence of blood supply. Scientific Reports, 2022, 12: 12006.

[153] He Mengying, Zhang Mengyao, Xu Tao, et al. Enhancing photodynamic immunotherapy by reprograming the immunosuppressive tumor microenvironment with hypoxia relief. Journal of Controlled Release, 2024, 368: 233-250.

[154] Lu Jun, Yang Li, Zhang Wei, et al. Photodynamic therapy for hypoxic solid tumors Via Mn-MOF as a photosensitizer. Chemical Communications, 2019, 55: 10792-10795.

[155] Zeng Jinyue, Zou Meizhen, Zhang Mingkang, et al. Π-Extended benzoporphyrin-based metal-organic framework for inhibition of tumor metastasis. ACS Nano, 2018, 12: 4630-4640.

[156] Chiu Huanmin, Chiou Wenyi, Hsu W J, et al. Salmonella alters heparanase expression and reduces tumor metastasis. International Journal of Medical Sciences, 2021, 18: 2981-2989.

[157] Barik G K, Sahay O, Mukhopadhyay A, et al. Fbxw2 suppresses breast tumorigenesis by targeting Akt-moesin-Skp2 axis. Cell Death & Disease, 2023, 14: 623.

[158] Frossi B, Antoniali G, Yu Kefei, et al. Endonuclease and redox activities of human apurinic/apyrimidinic endonuclease 1 have distinctive and essential functions in Iga class switch recombination. Journal of Biological Chemistry, 2019, 294: 5198-5207.

[159] Bowling F Z, Salazar C M, Bell J A, et al. Crystal structure of human Pld1 provides insight

into activation by Pi(4,5)P2 and Rhoa. Nature Chemical Biology, 2020, 16: 400-407.

[160] Zhang Long, Zhang Wan, Peng Hang, et al. Bioactive Cytomembrane@Poly(Citrate-Peptide)-Mirna365 nanoplatform with immune escape and homologous targeting for colon cancer therapy. Materials Today Bio, 2022, 15: 100294.

[161] Zhang Wenqing, Zhou Ronghui, Yang Yuting, et al. Aptamer-mediated synthesis of multifunctional nano-hydroxyapatite for active tumour bioimaging and treatment. Cell Proliferation, 2021, 54: e13105.

[162] Sun Jing, Long X, Li Rong, et al. Adriamycin loaded and folic acid coated Zn-MOF for tumor-targeted chemotherapy of cervical cancer. Journal of Biomaterials and Tissue Engineering, 2019, 9: 1535-1541.

[163] Nash G T, Luo Taokun, Lan Guangxu, et al. Nanoscale metal-organic layer isolates phthalocyanines for efficient mitochondria-targeted photodynamic therapy. Journal of the American Chemical Society, 2021, 143: 2194-2199.

[164] Alijani H, Noori A, Faridi N, et al. Aptamer-functionalized Fe_3O_4@MOF nanocarrier for targeted drug delivery and fluorescence imaging of the triple-negative mda-mb-231 breast cancer cells. Journal of Solid State Chemistry, 2020, 292: 121680.

第 5 章 MOFs 在抗癌领域的应用

癌症是一种极其顽固的疾病，具有极高的复发率、病死率和转移倾向，成为对人类健康构成重大威胁的元凶之一，每年夺去数百万人的生命[1]。尽管手术、放疗（RT）以及化疗是目前肿瘤治疗的常规手段，但这些治疗方式的精度不足、效率低下以及严重的副作用等问题，大大削弱了肿瘤[2]治疗的整体效果。至今，肿瘤治疗仍面临着诸多挑战，需要不断探索更加精准、高效且副作用较小的治疗方法，许多治疗方法，如化学动力学治疗（CDT）[3]、光热治疗[4]、饥饿治疗[5]、免疫治疗[6]，具有良好的低副作用、无创、方便操作，已被广泛开发提高抗癌效果满足日益增长的临床治疗的需求。然而，在癌症治疗的过程中，治疗策略的制定与实施依然面临着诸多严峻的挑战（表 5.1）。其中，化学动力学治疗（CDT）在肿瘤微环境（TME）中存在过氧化氢含量的限制[7]。光动力学治疗（PDT）显示光敏剂的稳定性较差，对 O_2 的依赖性较高[8]。因此，研究人员致力于开发合适的方法来提高肿瘤管理过程中的治疗效果。近几十年来，纳米材料的不断发展为改善各种肿瘤治疗方法的缺陷带来了新的思路（图 5.1）。因此，不同的先进材料被用于肿瘤治疗，以获得高效、精度突出、副作用可忽略不计的优点。随着纳米科学的快速进步，新兴材料的不断发展以及病理机制的深入探索，越来越多的生物技术纳米药物出现，成为癌症诊断和治疗的选择。这些纳米材料通常具有一种或多种电、光学、磁、pH、声学或其他属性。纳米药物或治疗性纳米材料所涉及的各种治疗方法（如化疗、光热治疗[9]、动态治疗和基因治疗[10]）在抗癌策略中也引起了广泛的关注。

表 5.1 癌症治疗方法的特点和机制概述

治疗策略	治疗原则	优势	局限性
放疗	射线辐射	深度渗透，局部治疗	副作用严重，易引起抗辐射性
化疗	化疗药物	应用范围广，疗效好	生物利用度低，副作用严重，易产生副作用
化学动力学治疗	芬顿型反应	外部能量和设备独立	多药耐药性以及治疗高度依赖于 H_2O_2 浓度
光热治疗	将近红外激光转化为超高温	特异性和低毒性	特异性低，光穿透有限

续表

治疗策略	治疗原则	优势	局限性
光动力学治疗	ROS/$^1O_2$产生光	可忽略的副作用低毒副作用，非侵入性和高选择性	光照深度有限，高度依赖 O_2 和特定装置
饥饿治疗	营养素消耗	非侵入性，低副作用	TME 缺氧加重，对 O_2 的依赖性高
免疫治疗	免疫反应	应用广泛范围	会产生炎症反应、复杂免疫抑制 TME 和非典型临床反应率

尽管刺激响应型 MOFs 在按需药物传递方面取得了快速发展，旨在减少癌症治疗期间的不良副作用，但仍有许多挑战亟待解决。例如，化疗过程中常见的多药耐药问题、单药治疗的局限性，以及肿瘤本身所具有的异质性和极其复杂的特性，这些都使得开发一种多功能的药物递送系统（DDS）成为迫切需求。为了满足这些复杂且多变的需求，多功能 DDS 应运而生并不断发展。其中，基于可控药物释放的调制器以及具有诊断和/或治疗功能的平台展现出了巨大的潜力。此外，多模式协同治疗策略，如光热治疗（PTT）[11]和光动力学治疗（PDT）[12]等，在癌症的治疗和诊断方面也表现出了显著的优势。这些技术的结合为癌症治疗提供了新的可能性，有望在未来取得更大的突破（图 5.2）。

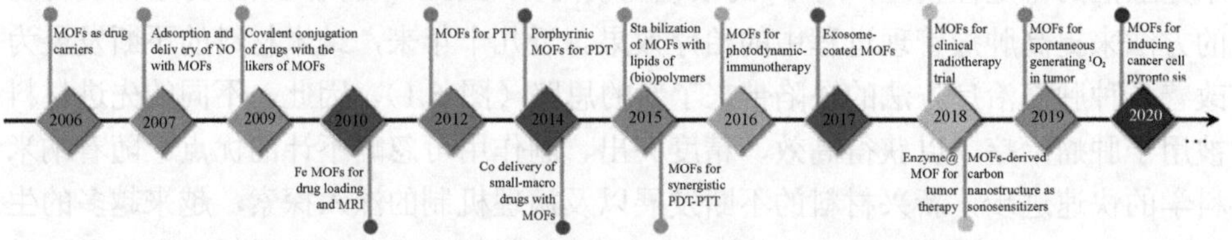

图 5.1　基于 MOFs 的抗肿瘤药物开发的简短时间线[13]

在设计用于癌症治疗的纳米材料时，必须综合考虑一系列关键因素以确保其有效性和安全性。这些关键因素包括：确保材料具备优异的稳定性，以应对复杂的生物环境；追求高药物转载量，以提高治疗效率；实现精确的靶向功能，确保药物能够精准到达肿瘤部位；实现药物的可控与持续释放，以优化治疗效果并减少副作用；同时，纳米材料还需具备良好的生物相容性和可降解性，以降低对生物体的潜在危害；此外，具备光、电、热、磁等物理、化学性质也是设计过程中不可或缺的考虑因素，这些性质能够拓展纳米材料在癌症治疗中的应用方式，并提升治疗效果。

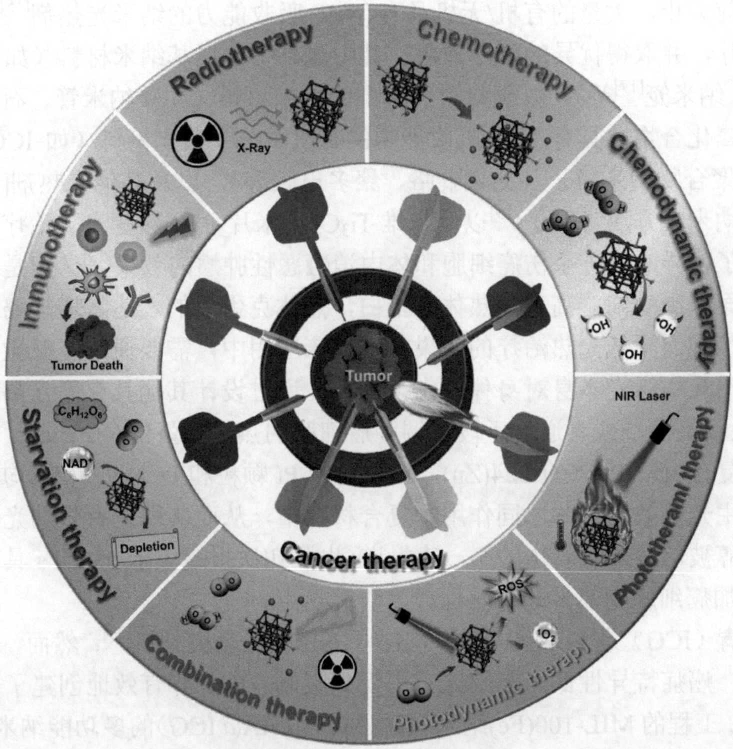

图 5.2　基于 MOFs 纳米载药平台的抗癌应用（包括 RT、化疗、CDT、PTT、PDT、饥饿治疗、免疫治疗和联合治疗）[14]

5.1 光热治疗

近年来，光热治疗（photothermal therapy，PTT）作为一种由红外光引发的治疗手段，备受瞩目。近红外光，其波长范围一般介于 650～900 nm，因具备出色的组织穿透深度和相对较低的组织吸收率，在癌症治疗中占有举足轻重的地位。它几乎不会对正常组织和细胞造成损伤，这一特性使得近红外光成为远程控制定点药物释放、热疗以及光动力学治疗等操作的理想选择。通过这些治疗方式，近红外光在癌症治疗中发挥着至关重要的作用。

当光热制剂受到特定波长（常在近红外范围内）的光激发后，这些制剂在回到基态的过程中，会释放振动能量进而产生热量。这种非辐射释放过程使得光能迅速转化为局部热量，从而利用高温有效杀死肿瘤细胞，为癌症治疗提供了新的途径。PTT 作为一种新兴的治疗方法，具有无创性、高选择性和组织穿透深度，能够通过光热剂将近红外激光转化为热疗，从而对肿瘤细胞进行光热消融。

到目前为止，大量的有机/无机具有 NIR 吸收能力的纳米光热剂[15]已被应用于光热治疗，并取得优异的治疗效果。其中包括贵金属基纳米材料（如 Au 纳米棒[16]、Au 纳米笼[17]等），以碳材料为基础的纳米材料（如碳纳米管、石墨烯等），过渡族金属化合物，具有 NIR 吸收能力的有机小分子荧光染料（如 ICG、IR825 等）[18]，聚合物纳米粒子（如聚吡咯、聚多巴胺等），以及其他光热剂[如黑磷，普鲁士蓝纳米立方块（PB）[19]以及二维 Ti_3C_2 纳米片等]。这些光热治疗剂在光热方面取得了优异的体外杀伤癌细胞和体内消融恶性肿瘤的效果[20]。但是，光热治疗通常会导致肿瘤细胞高表达热休克蛋白，热休克蛋白的表达将增加癌细胞对热量的耐受性进而降低光热治疗的效果。在临床应用中，需要通过降低激光功率来减轻光热治疗中激光本身对身体的伤害，因此通过设计其他具有高光热转换效率的 NIR 吸收光热剂或者通过手段下调肿瘤细胞的热休克蛋白[21]。Jiang 等合成了一种纳米复合材料 Pt-PCN-224(Zn)，该材料中 Pt 颗粒和 PCN-224(Zn)均能在光照条件下产生光热效应，并协同作用于复合材料中，从而实现更有效的光热效应。这是芳香醇被单线态氧氧化的第一次报道，也是初次报道关于 nMOFs 具有光热效应[22]以增加癌细胞副热量的敏感性成为研究热点。

靛花青（ICG）是 FDA 认可的近红外（NIR）区有机染料[23]；然而，由于 ICG 水溶性差，癌症特异性低，其临床应用受到限制。Cai 等有效地创建了一个基于 HA 和 ICG 工程的 MIL-100(Fe)纳米颗粒（MOF@HA@ICG）的多功能纳米平台[24]，用于成像引导的抗癌 PTT。合成的纳米颗粒具有较高的 ICG 负载率（40%）、良好的生物相容性、NIR 吸光度和光稳定性。Wu 等将光敏剂靛花青（ICG）和化疗药物阿霉素（Dox）逐步封装到 MIL-88 核和 ZIF-8 壳的纳米孔中，构建协同光热/光动力学/化疗纳米平台。除了有效的药物传递外，MIL-88 还可以作为一种纳米马达，将肿瘤微环境中过量的过氧化氢转化为足够的氧气，用于光动力学治疗[25]。

Zheng 等结合光敏四氯化物（TCPC）和 UiO-66 获得了一种新的材料 TCPC-UiO[26]。研究表明，Hf 具有较高的 X 射线衰减能力，TCPC 具有光敏性，其组合具有较高的光热转换效率、高稳定性和良好的生物相容性。此外，小鼠还表现出良好的抗癌活性和低毒性。除此之外，在红外光作用下，它还表现出优异的光热特性和 PTT 处理效率。Liang 等首次构建了肿瘤微环境激活的 FCSP@Dox MOFs，用于基于铁下垂的肿瘤化学动力学/光热/化学治疗。制备的 FCSP@Dox MOFs 可通过 EPR 效应积累在肿瘤部位，在过表达的 GSH 肿瘤微环境中由于 GSH 诱导的二硫键断裂而解体，从而通过抑制 GPX4 的表达诱导铁垂病。该系统可以通过氧化还原肿瘤微环境特异性激活，通过与铁依赖的 Fenton 反应协同诱导铁塌、低温和化疗，在体内外均获得良好的抗肿瘤作用（图 5.3）[27]。

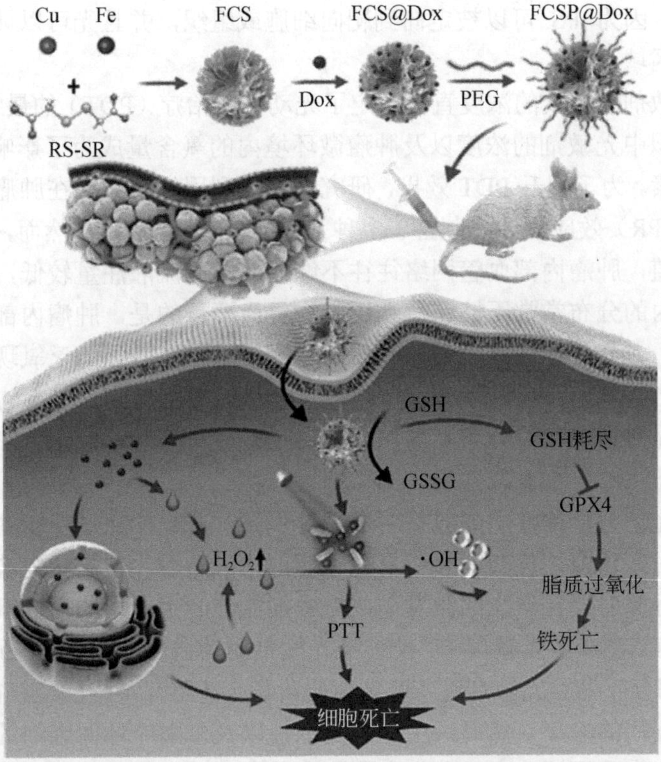

图 5.3　肿瘤微环境激活的 FCSP@Dox MOFs[27]

Deng 等首次合成了一种超薄（16.4 nm 厚）二茂铁基 MOFs（Zr-Fc MOF）纳米片，在无需额外药物的情况下 Zr-Fc MOF 纳米片在 808 nm 处的光热转化效率高达 53%，使其成为一种卓越的光热剂，并可用于协同光热治疗（PTT）和基于 Fenton 反应的化学动力学治疗（CDT），展现出其在癌症治疗领域的巨大潜力，而且可以促进过氧化氢转化为羟基自由基[28]。

5.2　光动力学治疗

光动力学治疗（PDT）[29]是一种于 2002 年提出的创新癌症治疗方法。这种方法的核心在于利用光照激活肿瘤部位富集的光敏剂（PS），这些光敏剂随后将周围的氧分子转化为高细胞毒性的活性氧（ROS），特别是单线态氧（$^1O_2$）。这些活性氧物质能够引发细胞凋亡和坏死，从而达到治疗癌症的目的。光动力学治疗为癌症治疗领域带来了新的希望和可能性。PDT 使用光敏剂与无害的可见光或近红外（NIR）光结合，产生活性氧（ROS）来杀死癌细胞或细菌。PDT 具有双重选择性的优势，因为 PS 可以被定向到靶向细胞或组织，并且光可以被物理地定向到受影响的区域。

由于光敏剂和氧气的浓度直接决定了光动力学治疗（PDT）的最终治疗效果，因此肿瘤组织中光敏剂的浓度以及肿瘤微环境内的氧含量成为了影响 PDT 疗效的关键性因素。为了提升 PDT 效果，研究人员已利用纳米颗粒在肿瘤部位增强渗透和滞留（EPR）效应设计并制备了多种 PS 装载的纳米制剂。然而，由于肿瘤细胞的恶性增殖，肿瘤内部血管网络往往不够发达，血流灌注量较低，这导致纳米颗粒在肿瘤内的分布常常不够充足和均匀。更为严重的是，肿瘤内部的氧气供应也严重不足，尤其是在远离血管的肿瘤细胞区域，存在明显的乏氧现象。这些因素都会严重影响 PDT 的治疗效果，使得这一疗法的应用受到了一定的限制。因此，如何解决肿瘤内纳米颗粒分布不均和氧气供应不足的问题，是提升 PDT 疗效的关键所在。

癌症的光动力学治疗主要包括两个关键步骤：首先，通过静脉注射或外敷方式将光敏剂引入体内，使其能够积累在肿瘤部位；随后，使用特定波长的激光对肿瘤区域进行照射。在激光的激发下，光敏剂能够产生大量的活性氧，从而有效地杀死癌细胞。在众多光敏剂中，卟啉类光敏剂因其显著的治疗效果而备受关注。目前，已有部分卟啉类光敏剂进入临床试验阶段，甚至有一些已经获得监管批准，应用于多种癌症的治疗。然而，由于其疏水性以及平面卟环间强烈的相互作用，这类光敏剂在生理环境中容易发生自聚沉现象。这种自聚沉不仅会降低光敏剂的活性，还会削弱其光吸收能力，进而减少单线态氧的产生效率，从而影响光动力学治疗的效果。为了解决这一问题，研究者们尝试将光敏剂包裹在纳米颗粒内部，以防止其降解和提前释放。通过纳米颗粒的被动或主动靶向作用，可以将光敏剂精确地递送到肿瘤病灶处。然而，简单地将光敏剂包裹在纳米颗粒内部可能会导致光敏剂之间距离过近，发生相互作用，引发自猝灭现象，降低光动力学治疗的效果。此外，活性 ROS 的扩散距离有限，通常为 20～200 nm，因此将光敏剂包裹进纳米材料内部可能会限制 ROS 的有效扩散，进而影响治疗效果。为了解决这些问题，研究者们需要进一步优化纳米颗粒的设计和制备方法，以确保光敏剂在纳米颗粒内部能够保持其活性和稳定性，同时实现高效的靶向递送和 ROS 扩散。这将有助于提升光动力学治疗的效果，为癌症治疗提供更为有效和安全的治疗手段。

卟啉作为一种多功能分子，在光动力学治疗（PDT）领域得到了广泛的开发和应用，其出色的光敏特性使得卟啉成为 PDT 中备受瞩目的光敏剂。更值得一提的是，卟啉基光敏剂因其独特的荧光特性，不仅能够实现高效的癌症治疗，还能在荧光成像引导下为治疗过程提供精确的定位和监测，因此被视为一种极具价值的治疗体系。这种荧光成像引导下的治疗方法为癌症的精准治疗提供了新的可能性，有助于提升治疗效果并降低副作用。2014 年，Lin 和同事报道了一种高效的 Hf-卟啉 NMOF PS，DBP-UiO，由 5,15-di(对苯并酸)卟啉（H_2DBP）配体和 Hf^{4+} 组成，通过 PDT 用于耐药头颈部癌症[30]。随后，他们设计并报道了第一个具有增强光物理性能和更高效的氯基 NMOF $^1O_2$ 生成能力，即 DBC-UiO，由 Hf^{4+}和卟啉衍生物（Me_2DBP）配体构建[31]。与 DBP-UiO 相比，氯基 DBC-UiO 显示出 13 nm 的红移，最低能量波段的消光系数提高了 11 倍，从而导致 $^1O_2$ 的产生增加了 3 倍，并增强了对 CT26 和 HT29 荷瘤小鼠的 PDT 效应。

最近，有研究报道卟啉基 MOFs 与 UCNPs 联合治疗 NIR 触发的 PDT，并增强了疗效[32]。Li 等报道了由 PCN-224 和 PVP 包被的镧系元素掺杂 UCNPs 组成的异源二聚体（UCMOFs），用于 NIR 引发的肿瘤治疗[33]。在 PCN-224 中，UCNPs 吸收 NIR 光并发射可见光，将能量转移到 TCPP 中，从而产生 $^1O_2$。结果表明，在 980 nm 激光照射下，UCMOFs 可以代替 PCN-224 产生细胞毒性 $^1O_2$。此外，体内抗肿瘤活性试验证实了 UCMOFs 具有显著的 NIR 触发的 PDT 能力。该策略使具有 NIR 光捕获功能的卟啉基 MOFs 能够应用于大型或深层肿瘤的近红外触发 PDT。他们还通过 DNA 介导的 PCN-224 与 UCNPs 的组装开发了 MOF-UCNP 核壳纳米颗粒。在 NIR 激光照射下，MOF-UCNP 纳米颗粒可以产生 $^1O_2$，产率随着 UCNPs 负荷的增加而增加。

为了解决 PDT 过程中 ROS 的缺氧和使用效率的问题，最近，Wang 等报道了一种超小型的卟啉-MOF 纳米点（MOF QDs），这种纳米点展现出了高活性氧（ROS）生成能力、高效的肿瘤积累特性以及在体内实现快速肾脏清除[34]的特点。这些优势不仅增强了光动力学治疗（PDT）的效果，还有效克服了长期毒性的问题，为癌症治疗提供了新的有力手段。Luo 等设计了一种高效的 PS、Hf 卟啉 nMOF 或 DBP-UiO[35]，用于治疗头颈部耐药癌症。他们发现 Hf(Ⅳ)-MOFs 表现出给药能力，并且与 PDT 药物联合使用可以避免 PS 的激发，从而导致自爆炸效应。此外，提高 Hf(Ⅳ)-MOFs 的孔隙率有助于 ROS 的扩散。采用 PDT 治疗小鼠模型头颈部癌症，DBP-UiO 的使用可完全根除老鼠的肿瘤，展示出优越的治疗效果。

光动力学治疗还可以与其他疗法相结合，Zhang 等提出了一种具有高稳定性和肿瘤靶向能力的多仿生 MOF 传递系统，用于肿瘤联合治疗、化疗和光动力学治疗（PDT）铁和卟啉集成纳米载体应促进联合治疗。FeTPt@CCM 被癌细胞内化后，可以通过 Fenton 反应以及 Fe^{3+}与谷胱甘肽和过氧化氢之间的氧化还原反应激活，生成羟基自由基和氧。因此，该纳米平台有效地诱导 Ferroptosis，提高了光动力学治疗的性能[36]。Liu 等提出了一种 PDT/化疗联合策略，将阿霉素（Dox）基卟啉 MOFs（PCN）表面组装 Dox 封装的沸石咪唑酯框架（ZIF-8）[37]。ZIF-8 的酸性 pH 反应降解促进了 Dox 在酸性肿瘤微环境中的释放。在光照射下，卟啉式 MOFs 可以捕获光，产生丰富的单线态氧，从而放大化疗/PDT 的治疗效果。

5.3 化　疗

生物分布薄弱是化疗的主要问题之一，因此需要高剂量进行治疗。在高频率的剂量下可以观察到一些副作用。这些情况需要新的和有效的药物递送系统。传统肿瘤治疗过程中，化疗方法因其不良副作用和低选择性一直受到挑战，这促使研究者们不断深入探索改进给药系统。在过去的几十年里，多种药物载体被开发出来，旨在提高抗癌药物的生物利用度和治疗效果。其中，无机载体如介孔二氧化硅纳米颗粒[38]、碳结构、氧化物和氮化物[39]等，以其独特的物理和化学性质在药物输送领域显示出巨大潜力。然而，尽管无机载体在药物传递方面取得了显著进展，但其生物分布、代谢机制和免疫原性等问题仍亟待进一步系统研究。相比之下，有机载体如聚合物、脂质体和树枝状大分子[40]等也广泛应用于药物递送系统，但往往面临低载量和早漏等挑战。因此，在开发新型药物载体时，需要综合考虑载体的生物相容性、药物释放动力学以及对肿瘤组织的靶向性等多个因素，以实现更高效、安全的癌症治疗。

随着癌症生物学领域的深入探索，近几十年来，众多抗癌制剂如小分子抑制剂、抗体、化疗药物以及核酸药物[41]等已陆续进入临床应用。然而，现行的治疗方法普遍面临非特异性分布和不佳的药代动力学等挑战，导致需使用大剂量药物，进而引发严重的毒副作用。为解决这些难题，纳米金属有机框架（NMOFs）作为一种极具潜力的纳米药物载体崭露头角。它们能够精准地将治疗药物递送至身体的特定病灶部位，并在相对较低的安全剂量下实现增强的治疗效果，从而显著改善了传统给药方式的局限。NMOFs 之所以具备这些优势，一方面是因为它们可以利用肿瘤部位特有的增强渗透和滞留（EPR）效应，实现药物在肿瘤部位的高效富集。另一方面，其巨大的比表面积、高孔隙率以及丰富的官能团，使得 NMOFs 能够有效装载并稳定药物分子，实现可控的药物释放。更为重要的是，通过对 NMOFs 进行靶向修饰，可以进一步提高药物在肿瘤部位的浓度，从而实现更为精准和高效的治疗效果。这种创新的给药方式有望为癌症治疗带来革命性的突破，为患者提供更加安全、有效的治疗选择。

Lin 和他的同事[42]报道了基于 UiO 的 NMOFs 平台，即 siRNA/UiO-Cis，用于联合传递顺铂和小干扰 RNA（siRNAs）治疗顺铂耐药卵巢癌。利用 UiO 固载顺铂前药和 UiO-Fs 的 Zr^{4+} 金属离子配位，进入 UiO 孔，将 siRNA 分别整合在 NMOFs 表面。这种共传递策略具有显著优势，它能够有效保护 siRNA 免受核酸酶的降解，进而促进 siRNA 从核内体中逃逸。通过这种方式，siRNA 能够成功沉默多个耐药基因，从而大幅增强顺铂的化疗效果，为癌症治疗提供了更为高效且可靠的方法。

除了将药物分子转化为前药并与金属离子配位，以生成 NMOFs/NCPs 直接应用于癌症治疗外，还有许多药物已成功实现在 NMOFs/NCPs 上的高效装载，并取得了显著的治疗效果。值得注意的是，药物的装载不仅可以在 NMOFs/NCPs 形成后进行，还可以在制备过程中就加入药物，从而实现药物的有效装载[43]。这种灵活的装载方式使得 NMOFs/NCPs 在药物传递和癌症治疗领域展现出巨大的应用潜力。

有研究将以四氧化三铁为核心壳纳米结构 Fe_3O_4@MOF 装载抗癌药物阿霉素（Dox）[44]，并将负载的纳米结构偶联到高荧光碳点（CDs）上。Fe_3O_4@MOF-DOX-CDs-Apt 纳米载体在遭遇过表达的癌细胞时，能特异性地触发其内部负载的释放机制。这一过程通过 pH 依赖性的控制释放动力学来实现，确保了药物传递的高效性，从而带来了更为显著的治疗效果。相较于正常的 HUVEC 细胞，这种刺激响应性的药物递送系统显示出卓越的选择性，使 Fe_3O_4@MOF-DOX-CDs-Apt 纳米载体能够精确地渗透到三阴性人乳腺癌细胞中。细胞毒性实验的结果进一步验证了这一点，纳米载体有效抑制了 MDA-MB-231 癌细胞的增殖，并在 24 小时内诱导了高达 77%的选择性凋亡率，而对正常 HUVEC 细胞的凋亡诱导率仅为不到 10%。这一发现为癌症治疗提供了新的策略，有望实现更为精准和高效的治疗效果。

Imaz 等成功合成了基于 Zn^{2+}和 1,4-双(咪唑-1-基甲基)苯的 Nano-MOFs，用于装载阿霉素和其他抗癌分子，其在细胞层面展现出卓越的抗肿瘤效果。在目前的药物装载方法中，研究者普遍先制备 Nano-MOFs，然后再将目标分子填充至其孔道内。然而，这种先制备后装载的方式往往导致目标药物的浪费。为了克服后装载所带来的药物浪费问题，Zou 等采用一步法原位技术，成功实现了 ZIF-8 的形成以及 DOX 的装载。研究结果显示，Dox 分子能够均匀地被封装在 ZIF-8 晶体中[45]。通过调整初始物质的相对含量，研究者还能够对 Dox 的装载量进行精确调控（图 5.4）。此外，该药物载体具备出色的 pH 响应药物释放性能，使得在细

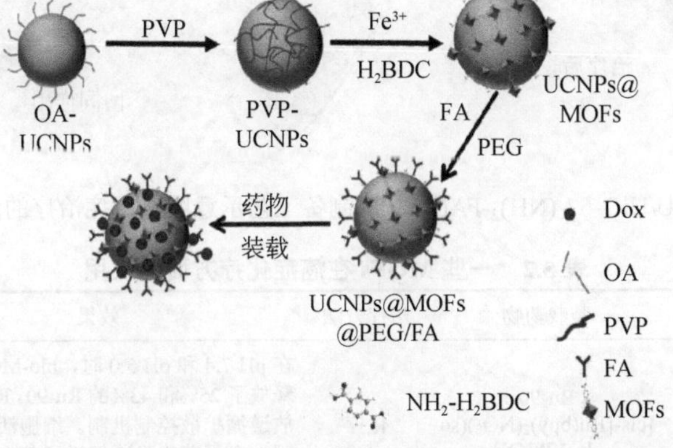

图 5.4　UCNPs@MIL-53@PEG/FA 治疗平台的形成工艺示意图[46]

胞水平的治疗效果明显优于自由态的 Dox。这一创新性的研究不仅提高了药物利用率，还为癌症治疗领域带来了新的突破。

Liu 等用顺铂前药将基于 Hf 的 MOFs 离子组装在牛血清白蛋白（BSA）稳定的二氧化锰（BM）纳米颗粒上[47]。用聚乙二醇（PEG）修饰后合成了 BM@NCP（DSP）-PEG 纳米颗粒。过氧化氢的分解代谢诱导原位 O_2 的产生，从而克服缺氧相关的 RT 抗性。Liu 等设计并构建了一种新的多功能药物载体 5-FU/UiO-67-$(NH)_2$-FAM/PMT（FUFP），纳米载体为 5-FU/UiO-67-$(NH)_2$，FAM 作为荧光成像剂，PMT 作为 FA 靶点和抗肿瘤药物（图 5.5）[48]。该药物载体具有多种功能，包括共载药物、靶向传递和荧光成像，以增强肿瘤治疗的抗肿瘤细胞毒性。MOFs 载药化疗这一策略未来可能对癌症治疗提供有效方案（表 5.2）。

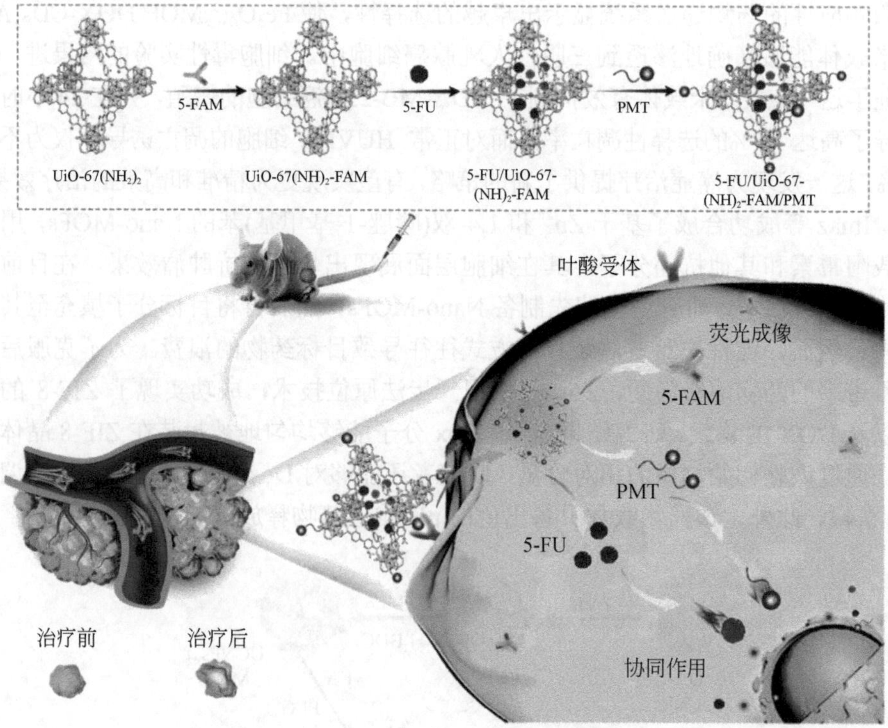

图 5.5　5-FU/UiO-67-$(NH)_2$-FAM/PMT 制备工艺示意图及癌症治疗的潜在机制[48]

表 5.2　一些 MOFs 在癌症化疗方面的应用

MOFs	装载药物	治疗方法	效果	参考文献
Bio-MOF-1	Ru-90 [cis-[Ru(bpy)$_2$(NO$_2$)(solv)](PF6)]	化疗	在 pH 7.4 和 pH 5.0 时，bio-MOF-1 分别释放了 25%和 43%的 Ru-90。Ru-90 的释放遵循扩散控制机制。细胞活力研究表明，金属药物阻断促进了其在细胞中的进入和可用性	[49]

续表

MOF	装载药物	治疗方法	效果	参考文献
Dox/Cel/ MOFs@Gel	阿霉素（Dox）和塞来昔布（Cel）	化疗	Cel 与阿霉素联合使用，以提高治疗效果。这款医疗平台展现出了出色的性能，其载药能力卓越，同时拥有稳定的 pH 响应释放特性。更为重要的是，该平台显著增强了体外对口腔癌细胞（包括 KB 和 SCC-9）的毒性作用，为口腔癌的治疗提供了新的有力工具	[50]
AuNC/Cam@MOF	喜树碱（Cam）	化疗	在该纳米探针中，NH_2-MIL-101（Fe）MOFs 可以作为药物纳米载体和猝灭剂，而且可为荧光成像和 ICP-MS 定量的双模式标签	[51]
CCM@N_3-bio-MOF-100/FA	姜黄素（CCM）	化疗	该材料具有良好的生物相容性和增强的药物释放能力在酸性条件下，与生理上的 pH 相比，显示出 pH 响应性药物的释放。CCM@N_3-bio-MOF-100-3D/FA 对 4T1 细胞具有良好的受体特异性靶向作用，在杀死癌细胞方面表现出显著的杀伤效果	[52]

5.4　放射治疗与声动力学治疗

5.4.1　放射治疗

放射治疗（radiotherapy，RT）作为一种常见的治疗方法已广泛引入癌症治疗。放射治疗[53]即利用 X 射线等电离辐射手段来杀灭癌细胞，是临床上常用的癌症治疗手段之一。目前，科研人员广泛探索含有高原子序数元素（如铋、金和稀土元素）的纳米颗粒，这些纳米颗粒能有效吸收高能量的电离辐射（如 X 射线），进而作为放射增敏剂来强化放疗对癌症的治疗效果。同样地，通过结合高原子序数金属离子与配体分子合成的 NMOFs/NCPs 也展现出了作为放疗增敏剂的潜力，能够进一步提升放疗的效果。这种创新性的材料设计为癌症放疗提供了新的可能性，有望在未来癌症治疗中发挥更加重要的作用。

Liu 等报道了 Hf^{4+}和 TCPP 通过溶剂热过程组装的 Hf-TCPP NMOFs，其中 TCPP 连接子表现出良好的 PDT 能力[54]。Hf^{4+}具有较强 X 射线衰减系数，可作为放射增敏剂增强放射治疗（RT）。由于重 Hf^{4+}节点增强了 ISC，Hf-TCPP NMOFs 在 661 nm 照射下非常有效地产生 $^1O_2$。在 4T1 乳腺荷瘤小鼠体内的结果显示，与单独 PDT 或 RT 相比，PDT 和 RT 与 Hf-TCPP NMOFs 协同显著抑制肿瘤生长，增强治疗疗效。

Wang 等合成了由 UiO 骨架构建的 Hf-MOF 和 Zr-MOF，并将其作为高效的 X 射线吸收器[55]。实验表明，Hf(Ⅳ)-MOF 和 Zr-MOF 均表现出显著的辐射发光，Hf(Ⅳ)-MOF 溶液混合 Zr-MOF 溶液的发光强度信号远高于含 Hf-MOF 或 Zr-MOF 的溶液。这些实验表明，基于这两种高 *Z* 元素的 MOFs 可以提高 X 射线的吸收，并在一定程度上提高了 RT 的治疗效率。

Zhou 等随后在这个方向上进行了类似的实验，他们在大气压下合成了 Hf(Ⅳ)-MOFs：UiO-66-NH_2（Hf），并将其作为治疗食管癌的放射增敏剂[56]。UiO-66-NH_2（Hf）是一种有效的 X 射线放射增敏剂，在体内和体外具有显著的 X 射线吸收能力，对食管鳞状细胞系 KYSE150 具有显著的放射增敏作用。通过对放射治疗的深入研究发现：当 DNA 损伤时，细胞进化出一种复杂的机制来维持基因组完整性，称为 DNA 损伤反应（DDR），它修复 DNA 以确保基因组的稳定性，这对肿瘤的治疗非常有害。因此，对 DDR 抑制剂的研究被认为可以对抗这一机制。Neufeld 等将 NMOFs 与 Hf 和配体 1,4-二羧苯（Hf-BDC）与两种 DDR 抑制药物塔拉唑帕布和布帕利布结合，得到复合纳米材料 TB@Hf-TB@HfBDC-PEG[57]。这种新设计的材料具有良好的药物释放控制效率，有效提高了 ROS 产生率、DNA 损伤和放射敏感性，使其成为避免 DDR 的有效治疗平台，为未来 DDR 抑制剂的开发提供了新的途径。

5.4.2 声动力学治疗

声动力学治疗（SDT）作为一种超声（US）作用下的声敏剂介导技术[58]，是一种很有前途的抗癌治疗方式，并正在成为一个前沿的跨学科研究领域。因为不同的生物效应引起其不同的强度和频率：低强度超声辐照可以产生非热生物效应治疗，超声空化作为刺激微泡加速治疗药物的深层肿瘤或实体肿瘤。此外，高强度 US 照射的生物学效应也可用于治疗。高强度聚焦超声照射通过诱导热疗可完成肿瘤消融，高强度脉冲 US 照射可实现肿瘤热分离。它整合了声增敏剂的局部细胞毒性和低强度的超声照射。SDT 与 PDT 类似，活性氧（ROS）是处理不可切除肿瘤的关键手段，特别是当增敏剂能够同时响应两种不同能源——即 PDT 的光刺激和 SDT 的 US 照射时（图 5.6）。SDT 不仅快速、无创，而且具有高度协作性，US 辐照的特点在于其低毒性和可重复性。尽管 PDT 已被证实在多种癌症治疗中具有显著效果，但光在穿透组织时的深度限制却大大阻碍了其在实体肿瘤、大型肿瘤或深部肿瘤中的应用。因此，SDT 作为一种创新的治疗方式，有望成为传统 PDT 的有力补充和替代方案，为肿瘤治疗开辟新的可能性。

在制备策略中，考虑将固有声增敏剂简单封装到 MOFs 材料中，以自然获得多模基声增敏剂。Zhang 等使用 Cu-MOF 载体负载 Ce，最后制造缺氧响应声增敏剂通过 UV-可见光谱定量后，Cu-MOF 中 Ce 的装载能力为 8.7%，缺氧条件下

Cu-MOF 释放的 Ce 比常氧状态下多，说明缺氧环境可以提高 Ce 的释放效率或 Ce 的低氧释放行为[59]。同样地，Huang 等将缺氧引发的 Cu-MOF 与偶氮引发剂 AIPH 混合使用，旨在获得肿瘤微环境（TME）响应的声增敏剂[60]。

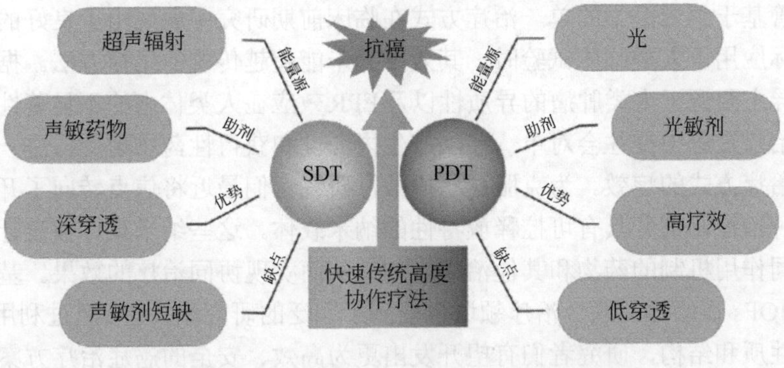

图 5.6 PDT 和 SDT 在癌症治疗中的主要比较

新型的声动力学治疗（SDT）和光动力学治疗（PDT）已经得到了很好的发展，并引起了许多研究者的关注。虽然这两种动力学治疗都是无创的、有效的肿瘤治疗方法，但它们都受到肿瘤缺氧环境的影响，所以降低了其治疗效率。为了解决这类治疗中的缺氧问题，Gao 等通过将合成的卟啉骨架掺杂具有过氧化氢酶样活性的铂纳米颗粒，合成了 Sm-TCPP-Pt/TPP 复合材料[61]。该材料具有独特的二维结构，可以促进活性氧的扩散，从而为 PDT 提供了一个富氧的治疗环境，克服了传统 PDT 的缺陷。Burger 等还利用铂纳米酶和含铬卟啉骨架制备了 PCN-224/Pt 复合材料，并加入阿霉素进行处理[62]。这种复合材料可以将内源性过氧化氢转化为氧气，从而减轻缺氧。

在声动力学治疗的基础上，除了传统的治疗方式外，研究人员还积极探索并开发出了微波治疗这一新兴手段。微波热疗凭借其操作简单、副作用轻微的显著优势，受到了广泛关注，被视为一种极具潜力的癌症治疗方法。在这方面，Meng 及其团队取得了令人瞩目的进展。他们成功报道了一种通过水热法制备的生物相容且可生物降解的 Mn^{2+}掺杂 Zr-MOF 纳米立方体（Mn-ZrMOF NCs）。这种纳米立方体不仅能够协同利用微波动态治疗（MDT）和微波热治疗（MTT）来共同抗击肿瘤，而且展现出了高达 28.7%的热转化效率。更为重要的是，Mn-ZrMOF NCs 能够在微波的刺激下产生具有细胞毒性的羟基自由基（·OH），这一发现为开发新型的微波响应癌症治疗平台奠定了坚实的基础。

5.5 联合治疗

尽管基于纳米医学的单一治疗方式在临床前期研究中展现出了良好的疗效，但在实际应用于人类临床试验时，其效率并未能超越传统的化疗方法。据报道，这种差异主要源于人类肿瘤的异质性以及EPR效应在人类体内的不显著性。此外，肿瘤在治疗过程中往往会对单一疗法产生内源性和获得性药物耐受，进一步削弱了单一治疗方式的疗效。为克服这些挑战，研究者们最近将焦点转向了开发能够装载多种治疗制剂并具有可控释放特性的纳米载体。这些纳米载体被设计为能够结合不同作用机制的药物和其他治疗制剂，从而实现协同治疗的效果。基于这一理念，MOFs 在癌症的联合治疗领域也受到了广泛的研究和关注。通过利用 MOFs 的独特性质和结构，研究者们有望开发出更为高效、安全的癌症治疗方案，为人类健康事业作出重要贡献。虽然目前各种治疗方法在疾病治疗中面临着一些困难，但个体治疗的疗效并不是绝对的。为了进一步提高疗效和冷肿瘤反应、克服耐药性等目的，联合治疗已成为抗肿瘤临床研究的突破。

5.5.1 光热治疗和放疗的联合

Liu 等精心设计了一种由铪离子与 4-羧基苯基(TCPP)共同构建的 NMOFs[54]。在这个独特的 Hf-TCPP NMOFs 中，TCPP 不仅扮演着光动力学治疗（PDT）中关键的光敏剂角色，而铪(Hf)离子则因其强大的 X 射线吸收能力成为放射治疗(RT)中的理想放射增敏剂。经过聚乙二醇（PEG）的表面修饰后，这些 NMOFs 在静脉注射后能够有效地在肿瘤部位富集，从而实现在活体水平上将 RT 与 PDT 有机结合，达到显著的抗肿瘤治疗效果。此外，还有报道提及了 Mn/Hf 金属有机框架在癌症治疗中的应用，特别是其光热治疗（PTT）与放疗的联合策略。这两种疗法展现出强烈的协同治疗效果，这是因为 PTT 产生的低温与高温能够刺激肿瘤部位的血流增加，进而改善肿瘤的乏氧状态。这一效应不仅有助于 PTT 本身的疗效提升，还进一步增强了 Hf 在放疗中的增敏能力，为癌症治疗提供了新的可能性。

5.5.2 光热治疗和化疗的联合

近年来，触发传递策略因其能有效解决药物过早释放的问题而备受关注。例如，Wang 等成功合成了具有优良分散性的核壳结构聚吡咯（PPy）@MIL-100（Fe）纳米颗粒[63]。这种药物递送系统在装载了 Dox 之后，展现出对 pH 和 NIR 的响应性药物释放特性，进而实现了针对癌细胞的协同化疗与光热治疗，为癌症治疗领域带来了新的突破（图 5.7）。

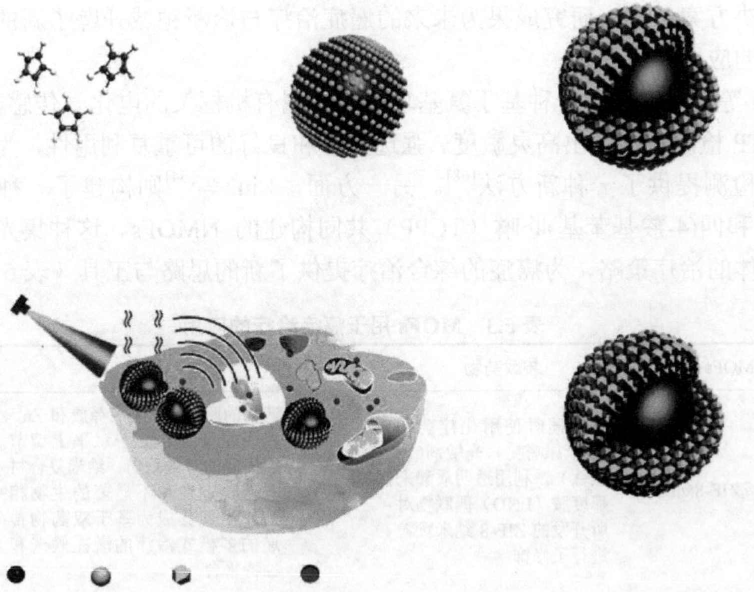

图 5.7　PPy@MIL-100（Fe）作为 pH/NIR 响应性药物载体，用于双模式成像和协同化疗–光热治疗

5.5.3　光动力学治疗和化疗的联合

在合适的配位溶剂作用下，含铁的 NMOFs 能够形成具备高自旋超顺磁性的复合物。这一特性使得这些复合物在磁性靶向的作用下能够精准地递送至癌细胞中，为癌症的磁靶向治疗提供了有力的工具。有研究表明，以铈盐和 2-甲基咪唑为原料，成功克服了 Ce 在高浓度时易发生的自猝灭现象，实现了非均相负载的高浓度 Ce 的光热和光动力学治疗[64]。此外，据报道，MOFs 载药能够响应肿瘤微环境中的过氧化氢，产生 O_2，从而有效改善肿瘤组织的乏氧状态。在进一步修饰聚乙二醇后，成功实现了三模态成像引导的化疗与光热光动力学联合治疗[65]，为癌症治疗领域带来了新的突破。

基于 DNA 功能化的纳米金属有机框架材料，有研究成功开发了一种创新的药物递送和光动力学治疗策略，能够精准追踪并靶向 A549 细胞。这一系统能够有效地将化疗药物 Dox 以及含有光敏剂的金属有机框架结构运送到肿瘤细胞内部[66]。实验数据表明，这种联合给药平台（化疗与光动力学治疗相结合）的疗效显著优于单一构建的给药平台，为癌症治疗提供了更高效、精准的治疗手段。值得一提的是，PCN-224-DNA-DOX 不仅具有优异的治疗效果，还能够作为肿瘤细胞检测的探针[67]。通过整合共给药、光动力学治疗以及诊断功能，实现了基于 PCN-224-DNA-Dox 的诊断–治疗一体化平台，为肿瘤的综合治疗与监测提供了全新的解决方案。这一研究成果为未来的癌症治疗与诊断领域开辟了新的道路，具有广阔的应用前景。

Shi 等巧妙设计了一种基于氨基功能化金属有机框架的电化学传感器，该传感器在 ATP 检测中展现出高灵敏度、强选择性和良好的可重复利用性，为生物分子的精确检测提供了一种新方法[68]。另一方面，Liu 等[54]则构建了一种由铪离子（Hf^{4+}）和四(4-羧基苯基)卟啉（TCPP）共同构建的 NMOFs。这种集光动力与放疗于一体的治疗策略，为癌症的综合治疗提供了新的思路与工具（表 5.3）。

表 5.3　MOFs 用于癌症治疗的应用

MOFs	装载药物	治疗方法	效果	参考文献
LND-HA@ZIF-8@Lf-TC	通过同时使用化疗药物（5-氟尿嘧啶）、光敏剂（二氯钛），利用透明质酸来那度胺（LND）偶联物对所开发的 ZIF-8 纳米颗粒进行了修饰		药物的两步 pH 响应释放和 Zn^{2+} 离子在细胞和分子水平上调节癌细胞的有效杀伤。纳米复合材料的稳定性和不显著的生物相互作用使其成为基于双药物传递的多模式治疗的潜在候选和独特平台	[69]
PEG-FA-NH_2-Fe-BDC	Dox		功能化 MOF 的胶体稳定性有所提高。Dox 的封装效率约为 97%，而封装能力约为 14.5%（质量分数）。此外，在不同的 pH 值（5.3 和 7.4）下，研究了有无低频超声（LFUS，40 kHz）的体外释放谱	[70]
PEG-PCN（PL-PEG-PCN）	哌隆胺（PL）	超声动力学治疗	PEG-PCN（PL-PEG-PCN）在生物介质中均表现出较高的胶体稳定性。促氧化载药卟啉啉 MOFs 是生物相容性和有效的肿瘤靶向化疗-超声动力学联合治疗的超声增敏剂	[71]
UCNPs@MIL-53@PEG/FA	阿霉素，荧光纳米粒子（UCNPs）		MIL-53 通过原位生长覆盖在 UCNPs 表面，具有较高的载药能力，进一步移植 FA 使系统能够定位病变 UCNPs@MIL-53@PEG/FA 的成功制备使得构建一个集成生物成像和药物传递的双功能治疗系统成为可能	[72]
IR820@ZIF-8 and（R837+1 MT）@ZIF-8	光热剂 IR820、辅助剂咪喹莫特（R837）和免疫调节剂 1-甲基-D-色氨酸（1 MT）	PTT	HA/IR820@ZIF-8 和 MAN/（R837+1 MT）@ZIF-8 的混合物可产生良好的系统免疫反应和免疫记忆效应，以对抗预先存在的转移性肿瘤和再挑战性肿瘤	[73]

续表

MOFs	装载药物	治疗方法	效果	参考文献
Cu-MOF		针对癌细胞的抗有丝分裂药物	具有内在蛋白酶样活性的 Cu-MOF 也能水解细胞骨架蛋白（F-actin）。Cu-MOFs 具有较强的抗肿瘤活性，至少是青蒿琥酯和奥沙利铂对 SKOV3 的 3 倍	[74]
MA-HfMOF-PFC Ni-Zn	光敏剂	PDT	光敏剂通过并入金属有机框架（MOFs）被送到癌细胞系进行 PDT。可以积极靶向三种癌细胞系这种氯基 MOFs 的抗癌能力比卟啉基 MOFs 高数倍	[75]
hM@ZMDF	阿霉素（Dox）、氧化锰（MnO_x）纳米颗粒和叶酸（FA）		Dox 作为化疗药物和荧光显像剂具有双重作用。此外，释放的 Fe^{3+} 可以介导 Fenton 反应，在高谷胱甘肽存在下，在细胞内产生有毒的羟基自由基	[76]
Cu-GA NMOF	没食子酸（GA）	PDT	与游离 GA 相比，这些晶体纳米材料可以被 Panc-1 细胞有效地吸收。细胞活力试验证实了 Cu-GA 通过产生 ROS 诱导的细胞毒性，半抑制浓度为 50 μg/mL	[77]
MSN-OVA@MOF	抗原提呈细胞（APCs）对肿瘤相关抗原（TAAs）		pH 响应性 MOFs 降解发生在酸性环境下，并在内切/溶酶体存在下释放抗原	[78]
F127-DOX-MIL	盐酸阿霉素	化疗和化疗动力学	纳米 MOFs 在 Dox 碱性水溶液中作为纳米海绵，负载能力高达 24.5%（质量分数），而负载效率高达 98%	[79]
RBCm@Ag-MOFs/PFK15 （A-RAMP）	肿瘤有氧糖酵解抑制剂 PFK15 （P）		A-RAMP 可以准确靶向肿瘤细胞，重编程有氧糖酵解，Ag^+ 和 PFK 协同抗肿瘤作用。通过干扰糖酵解和细胞凋亡介导的淋巴瘤的靶向治疗	[80]
Zr-Fc MOF		PTT/CDT	不仅是一种优良的光热剂，在 808 nm 处的光热转化效率为 53%，而且可以促进过氧化氢转化为羟基自由基（•OH）。因此，通过这种组合效应，在体外和体内都取得了良好的治疗效果	[28]
miR-34a-m@ZIF-8	MicroRNA（miRNA）	CDT	miR-34a-m@ZIF-8 复合物显示了有效的细胞摄取和溶酶体刺激应答的 miRNA 释放。Zn^{2+} 触发活性氧的产生，从而诱导肿瘤细胞凋亡	[81]

参 考 文 献

[1] 周鑫驰. 基于金属离子掺杂碳点的比色阵列传感器的生物硫醇检测新方法研究及其疾病诊断应用. 南京: 南京邮电大学, 2023.

[2] 张淑群, 马兴聪, 孙诗雨, 等. Car-T 细胞免疫疗法在实体瘤中的研究进展. 西南医科大学学报, 2024, 47: 98-103.

[3] 陈艾红, 张子文, 周诺馨, 等. 二氧化锰纳米复合材料的合成及其在肿瘤治疗中的应用进展. 山东化工, 2023,52: 78-81.

[4] Yang Kai, Zhang Shuai, Zhang Guoxin, et al. Graphene in mice: Ultrahigh *in vivo* tumor uptake and efficient photothermal therapy. Nano Letters, 2010, 10: 3318-3323.

[5] 李铭莹, 林霖, 王岩, 等. 人参皂苷抗肿瘤机制及其纳米药物递送系统的研究进展. 中草药, 2024,55: 688-696.

[6] Pardoll D M. The blockade of immune checkpoints in cancer immunotherapy. Nature Reviews Cancer, 2012, 12: 252-264.

[7] Lin Lisen, Song Jibin, Song Liang, et al. Simultaneous fenton-like ion delivery and glutathione depletion by Mno^{2+}-based nanoagent to enhance chemodynamic therapy. Angewandte Chemie International Edition, 2018, 57: 4902-4906.

[8] Nyman E S, Hynninen P H. Research advances in the use of tetrapyrrolic photosensitizers for photodynamic therapy. Journal of Photochemistry and Photobiology B, 2004, 73: 1-28.

[9] Tu Xiaolong, Wang Lina, Cao Yuhua. et al. Efficient cancer ablation by combined photothermal and enhanced chemo-therapy based on carbon nanoparticles/doxorubicin@SiO_2 nanocomposites. Carbon, 2016, 97: 35-44.

[10] Xin Yong, Huang Min, Guo Wenwen, et al. Nano-based delivery of rnai in cancer therapy. Molecular Cancer, 2017, 16: 134.

[11] 廖宇思, 梁剑箫, 温转, 等. 生物医用高分子材料细胞膜表面功能化的策略与应用, 高分子学报, 2024, 55: 553-572.

[12] Ethirajan M, Chen Yihui, Joshi P, et al. The role of porphyrin chemistry in tumor imaging and photodynamic therapy. Chemical Society Reviews, 2011, 40: 340-362.

[13] Gao Peng, Chen Yuanyuan, Pan Wei, et al. Antitumor agents based on metal-organic frameworks. Angewandte Chemie International Edition, 2021, 60: 16763-16776.

[14] Yang Jie, Dai Dihua, Zhang Xi, et al. Multifunctional metal-organic framework (MOF)-based nanoplatforms for cancer therapy: From single to combination therapy. Theranostics, 2023, 13: 295-323.

[15] Li Xiaozhen, Liu Lu, Li Shengliang, et al. Biodegradable ii-conjugated oligomer nanoparticles with high photothermal conversion efficiency for cancer theranostics. ACS Nano, 2019, 13: 12901-12911.

[16] Cui Xinyu, Li Minghui, Tong Lei, et al. High aspect ratio plasmonic Au/Ag nanorods-mediated Nir-Ii photothermally enhanced nanozyme catalytic cancer therapy. Colloids and Surfaces B: Biointerfaces, 2023, 223: 113168.

[17] Pakravan A, Salehi R, Mahkam M. Comparison study on the effect of gold nanoparticles shape in the forms of star, hallow, cage, rods, and Si-Au and Fe-Au core-shell on photothermal cancer treatment. Photodiagnosis and Photodynamic Therapy, 2021, 33: 102144.

[18] Schaafsma B E, Mieog J S D, Hutteman M, et al. The clinical use of indocyanine green as a near-infrared fluorescent contrast agent for image-guided oncologic surgery. Journal of Surgical Oncology, 2011, 104: 323-332.

[19] Wang Meng, Li Baolong, Du Yu, et al. Fluorescence imaging-guided cancer photothermal therapy using polydopamine and graphene quantum dot-capped prussian blue nanocubes. RSC Advances, 2021, 11: 8420-8429.

[20] Bai Zhiqiang, Zhao Lu, Feng Haidi, et al. Aptamer modified Ti_3C_2 nanosheets application in smart targeted photothermal therapy for cancer. Cancer Nanotechnology, 2023, 14: 35.

[21] Zhang Xinhao, Xue Shanshan, Pan Wei, et al. A heat shock protein-inhibiting molecular photothermal agent for mild-temperature photothermal therapy. Chemical Communications, 2023, 59: 235-238.

[22] Mohammed M R S, Ahmad V, Ahmad A, et al. Prospective of nanoscale metal organic frameworks [NMOFs] for cancer therapy. Seminars in Cancer Biology, 2021, 69: 129-139.

[23] Dai Qixuan, Ren En, Xu Dazhuang, et al. Indocyanine green-based nanodrugs: A portfolio strategy for precision medicine. Progress in Natural Science: Materials International, 2020, 30: 577-588.

[24] Cai Wen, Gao Haiyan, Chu Chengchao, et al. Engineering phototheranostic nanoscale metal-organic frameworks for multimodal imaging-guided cancer therapy. ACS Applied Materials & Interfaces, 2017, 9: 2040-2051.

[25] Wu Biyuan, Fu Jintao, Zhou Yixian, et al. Tailored core-shell dual metal-organic frameworks as a versatile nanomotor for effective synergistic antitumor therapy. Acta Pharmaceutica Sinica B, 2020, 10: 2198-2211.

[26] Zheng Xiaohua, Wang Lei, Liu Ming, et al. Nanoscale mixed-component metal-organic frameworks with photosensitizer spatial-arrangement-dependent photochemistry for multimodal-imaging-guided photothermal therapy. Chemistry of Materials, 2018, 30: 6867.

[27] Liang Yu, Zhang Li, Peng Chao, et al. Tumor microenvironments self-activated nanoscale

metal-organic frameworks for ferroptosis based cancer chemodynamic/photothermal/chemo therapy. Acta Pharmaceutica Sinica B, 2021, 11: 3231-3243.

[28] Deng Zheng, Fang Chao, Ma Xu, et al. One stone two birds: Zr-Fc metal-organic framework nanosheet for synergistic photothermal and chemodynamic cancer therapy. ACS Applied Materials & Interfaces, 2020, 12: 20321-20330.

[29] Aksel M, Bozkurt-Girit O, Bilgin M D. Pheophorbide a-mediated sonodynamic, photodynamic and sonophotodynamic therapies against prostate cancer. Photodiagnosis and Photodynamic Therapy, 2020, 31: 101909.

[30] Lu Kuangda, He Chunbai, Lin Wenbin. Nanoscale metal-organic framework for highly effective photodynamic therapy of resistant head and neck cancer. Journal of the American Chemical Society, 2014, 136: 16712-16715.

[31] Lu Kuangda, He Chunbai, Lin Wenbin. A chlorin-based nanoscale metal-organic framework for photodynamic therapy of colon cancers. Journal of the American Chemical Society, 2015, 137: 7600-7603.

[32] Li Yite, Zhou Junli, Chen Yuannan, et al. Near-infrared light-boosted photodynamic-immunotherapy based on sulfonated metal-organic framework nanospindle. Chemical Engineering Journal, 2022, 437: 135370.

[33] Li Yifan, Di Zhenghan, Gao Jinhong, et al. Heterodimers made of upconversion nanoparticles and metal-organic frameworks. Journal of the American Chemical Society, 2017, 139: 13804-13810.

[34] Wang Huan, Yu Dongqin, Fang Jiao, et al. Renal-clearable porphyrinic metal-organic framework nanodots for enhanced photodynamic therapy. ACS Nano, 2019, 13: 9206-9217.

[35] Luo Taokun, Fan Yingjie, Mao Jianming, et al. Dimensional reduction enhances photodynamic therapy of metal-organic nanophotosensitizers. Journal of the American Chemical Society, 2012, 144: 5241-5246.

[36] Zhang Qingfei, Kuang Gaizhen, Wang Hanbing, et al. Multi-bioinspired MOF delivery systems from microfluidics for tumor multimodal therapy. Advanced Science, 2023, 10: 2303818.

[37] Liu Bei, Liu Zechao, Lu Xijian, et al. Controllable growth of drug-encapsulated metal-organic framework (MOF) on porphyrinic MOF for pdt/chemo-combined therapy. Materials & Design, 2023, 228: 111861.

[38] Yue Juan, Luo Shizhong, Lu Mengmeng, et al. A comparison of mesoporous silica nanoparticles and mesoporous organosilica nanoparticles as drug vehicles for cancer therapy. Chemical Biology & Drug Design, 2018, 92: 1435-1444.

[39] Huang Junhao, Marcus K, Tian Xinxin, et al. Fundamental structural and electronic understanding of palladium catalysts on nitride and oxide supports. Angewandte Chemie

International Edition, 2024, 63: e202400174.

[40] Fernandes G, Pandey A, Kulkarni S, et al. Supramolecular dendrimers based novel platforms for effective oral delivery of therapeutic moieties. Journal of Drug Delivery Science and Technology, 2021, 64: 102647.

[41] Wang Shutao, Zhang Muxin, Liang Di, et al. Molecular design and anticancer activities of small-molecule monopolar spindle 1 inhibitors: A medicinal chemistry perspective. European Journal of Medicinal Chemistry, 2019, 175: 247-268.

[42] He Chunbai, Lu Kuangda, Liu Demin, et al. Nanoscale metal-organic frameworks for the Co-delivery of cisplatin and pooled sirnas to enhance therapeutic efficacy in drug-resistant ovarian cancer cells. Journal of the American Chemical Society, 2014, 136: 5181-5184.

[43] Kumeria T. Advances on porous nanomaterials for biomedical application (drug delivery, sensing, and tissue engineering). ACS Biomaterials Science & Engineering, 2022, 8: 4025-4027.

[44] Xue Zhongbo, Zhu Mengyao, Dong Yuze, et al. An integrated targeting drug delivery system based on the hybridization of graphdiyne and MOFs for visualized cancer therapy. Nanoscale, 2019, 11: 11709-11718.

[45] Jing Ziwei, Wang Xiaohui, Li Na, et al. Ultrasound-guided percutaneous metal-organic frameworks based codelivery system of doxorubicin/acetazolamide for hepatocellular carcinoma therapy. Clinical and Translational Medicine, 2021, 11: e600.

[46] Cong Hailin, Jia Feifei, Wang Song, et al. Core–shell upconversion nanoparticle@metal-organic framework nanoprobes for targeting and drug delivery. Integrated Ferroelectrics, 2020, 206: 66-78.

[47] Liu Jingjing, Chen Qian, Zhu Wenwen, et al. Nanoscale-coordination-polymer-shelled manganese dioxide composite nanoparticles: A multistage redox/pH/H_2O_2-responsive cancer theranostic nanoplatform. Advanced Functional Materials, 2017, 27: 1605926.

[48] Liu Weicong, Pan Ying, Zhong Yingtao, et al. A multifunctional aminated UiO-67 metal-organic framework for enhancing antitumor cytotoxicity through bimodal drug delivery. Chemical Engineering Journal, 2021, 412: 127899.

[49] Armando R A M, Abuçafy M P, Graminha A E, et al. Ru-90@Bio-MOF-1: A ruthenium(II) metallodrug occluded in porous Zn-based MOF as a strategy to develop anticancer agents. Journal of Solid State Chemistry, 2021, 297.

[50] Tan Guozhu, Zhong Yingtao, Yang Linlin, et al. A multifunctional MOF-based nanohybrid as injectable implant platform for drug synergistic oral cancer therapy. Chemical Engineering Journal, 2020, 390: 124446.

[51] Yin Xiao, Yang Bin, Chen Beibei, et al. Multifunctional gold nanocluster decorated

metal-organic framework for real-time monitoring of targeted drug delivery and quantitative evaluation of cellular therapeutic response. Analytical Chemistry, 2019, 91: 10596-10603.

[52] Alves R C, Schulte Z M, Luiz M T, et al. Breast cancer targeting of a drug delivery system through postsynthetic modification of curcumin@N_3-Bio-MOF-100 via click chemistry. Inorganic Chemistry, 2021, 60: 11739-11744.

[53] Liu Jingjing, Hu Fang, Wu Min, et al. Bioorthogonal coordination polymer nanoparticles with aggregation-induced emission for deep tumor-penetrating radio-and radiodynamic therapy. Advanced Materials, 2021, 33: 2007888.

[54] Liu Jingjing, Yang Yu, Zhu Wenwen, et al. Nanoscale metal-organic frameworks for combined photodynamic & radiation therapy in cancer treatment. Biomaterials, 2016, 97: 1-9.

[55] Chen Dashu, Xing Hongzhu, Wang Chungang, et al. Highly efficient visible-light-driven CO_2 reduction to formate by a new anthracene-based zirconium MOF *via* dual catalytic routes. Journal of Materials Chemistry A, 2016, 4: 2657-2662.

[56] Zhou Wei, Liu Zhulong, Wang Nana, et al. Hafnium-based metal-organic framework nanoparticles as a radiosensitizer to improve radiotherapy efficacy in esophageal cancer. ACS Omega, 2022, 7: 12021-12029.

[57] Neufeld M J, DuRoss A N, Landry M R, et al. Co-delivery of parp and Pi3k inhibitors by nanoscale metal-organic frameworks for enhanced tumor chemoradiation. Nano Research, 2019, 12: 3003-3017.

[58] Schaab L, Ferry Y, Ozdas M, et al. Exth-49 focused ultrasound and 5-Ala mediated elimination of diffuse midline glioma. Neuro-Oncology, 2021, 23: vi174.

[59] Zhang Kai, Meng Xiangdan, Yang Zhou, et al. Enhanced cancer therapy by hypoxia-responsive copper metal-organic frameworks nanosystem. Biomaterials, 2020, 258: 120278.

[60] Sun Yu, Cao Jing, Wang Xue, et al. Hypoxia-adapted sono-chemodynamic treatment of orthotopic pancreatic carcinoma using copper metal-organic frameworks loaded with an ultrasound-induced free radical initiator. ACS Applied Materials & Interfaces, 2021, 13: 38114-38126.

[61] Gao Zhiguo, Li Yaojia, Zhang Yu, et al. Biomimetic platinum nanozyme immobilized on 2d metal-organic frameworks for mitochondrion-targeting and oxygen self-supply photodynamic therapy. ACS Applied Materials & Interfaces, 2020, 12: 1963-1972.

[62] Burger T, Winkler C, Dalfen I, et al. Porphyrin based metal-organic frameworks: Highly sensitive materials for optical sensing of oxygen in gas phase. Journal of Materials Chemistry C, 2021, 9: 17099-17112.

[63] Chen Xiangjun, Zhang Manjie, Li Shengnan, et al. Facile synthesis of polypyrrole@ metal-organic framework core-shell nanocomposites for dual-mode imaging and synergistic

chemo-photothermal therapy of cancer cells. Journal of Materials Chemistry B, 2017, 5: 1772-1778.

[64] Wang Menglin, Zhai Yinglei, Ye Hao, et al. High Co-loading capacity and stimuli-responsive release based on cascade reaction of self-destructive polymer for improved chemo-photodynamic therapy. ACS Nano, 2019, 13: 7010-7023.

[65] Yang Hong Yu, Moon-Sun J, Li Yi, et al. pH-responsive dynamically cross-linked nanogels with effective endo-lysosomal escape for synergetic cancer therapy based on intracellular Co-delivery of photosensitizers and proteins. Colloids and Surfaces B: Biointerfaces, 2022, 217: 112638.

[66] Mukherjee P, Guha S, Das G, et al. Nir light-activated upconversion pop nanofiber composite; an effective carrier for targeted photodynamic therapy and drug delivery. Journal of Photochemistry and Photobiology A: Chemistry, 2023, 443: 114907.

[67] Feng Haidi, Zhao Lu, Bai Zhiqiang, et al. Aptamer modified Zr-based porphyrinic nanoscale metal-organic frameworks for active-targeted chemo-photodynamic therapy of tumors. RSC Advances, 2023, 13: 11215-11224.

[68] Shi Pengfei, Zhang Yuanchao, Yu Zhaopeng, et al. Label-free electrochemical detection of atp based on amino-functionalized metal-organic framework. Scientific Reports, 2017, 7: 6500.

[69] Pandey A, Kulkarni S, Vincent A P, et al. Hyaluronic acid-drug conjugate modified core-shell MOFs as pH responsive nanoplatform for multimodal therapy of glioblastoma. International Journal of Pharmaceutics, 2020, 588: 119735.

[70] Ahmed A, Karami A, Sabouni R, et al. pH and ultrasound dual-responsive drug delivery system based on peg-folate-functionalized iron-based metal-organic framework for targeted doxorubicin delivery. Collodis and Surfaces A—Physicochemical and Engineering Aspects, 2021, 626: 127062.

[71] Hoang Q T, Kim M, Kim B C, et al. Pro-oxidant drug-loaded porphyrinic zirconium metal-organic-frameworks for cancer-specific sonodynamic therapy. Collodis and Surfaces B-Biointerfaces, 2022, 209: 112189.

[72] Li Yantao, Tang Jinglong, He Liangcan, et al. Core-shell upconversion nanoparticle@metal-organic framework nanoprobes for luminescent/magnetic dual-mode targeted imaging. Advanced Materials, 2015, 27: 4075-4080.

[73] Zhang Huiyuan, Zhang Jing, Li Qian, et al. Site-specific MOF-based immunotherapeutic nanoplatforms via synergistic tumor cells-targeted treatment and dendritic cells-targeted immunomodulation. Biomaterals, 2020, 245: 119983.

[74] Chen Daomei, Li Bin, Jiang Liang, et al. Pristine Cu-MOF induces mitotic catastrophe and alterations of gene expression and cytoskeleton in ovarian cancer cells. ACS Applied Bio

Materials, 2020, 3: 4081-4094.

[75] Sakamaki Y, Ozdemir J, Heidrick Z, et al. A bioconjugated chlorin-based metal-organic framework for targeted photodynamic therapy of triple negative breast and pancreatic cancers. ACS Applied Bio Materials, 2021, 4: 1432-1440.

[76] Zeng Xiaoli, Chen Bin, Song Yibo, et al. Fabrication of versatile hollow metal-organic framework nanoplatforms for folate-targeted and combined cancer imaging and therapy. ACS Applied Bio Materials, 2021, 4: 6417-6429.

[77] Sharma S, Mittal D, Verma A K, et al. Copper-gallic acid nanoscale metal-organic framework for combined drug delivery and photodynamic therapy. ACS Applied Bio Materials, 2019, 2: 2092-2101.

[78] Duan Fei, June W, Li Zhaoxi, et al. pH-responsive metal-organic framework-coated mesoporous silica nanoparticles for immunotherapy. ACS Applied Bio Materials, 2021, 4: 13398-13404.

[79] Yao Lijia, Tang Ying, Cao Wenqian, et al. Highly efficient encapsulation of doxorubicin hydrochloride in metal-organic frameworks for synergistic chemotherapy and chemodynamic therapy. ACS Biomaterials Science & Engineering, 2021, 7: 4999-5006.

[80] Zhao Qiangqiang, Li Jian, Wu Bin, et al. Smart biomimetic nanocomposites mediate mitochondrial outcome through aerobic glycolysis reprogramming: A promising treatment for lymphoma. ACS Applied Materials & Interfaces, 2020, 12: 22687-22701.

[81] Zhao Huaixin, Li Taotao, Yao Chi, et al. Dual roles of metal-organic frameworks as nanocarriers for miRNA delivery and adjuvants for chemodynamic therapy. ACS Applied Materials & Interfaces, 2021, 13: 6034-6042.

第 6 章 MOFs 在药物递送领域应用的多元化

在药物递送领域，金属有机框架（MOFs）的应用日益受到重视，特别是在提高药物疗效和降低副作用方面展现出独特优势。MOFs 凭借其高度可调的孔隙性、较大的比表面积以及对外部刺激的响应性，为药物递送系统（DDS）带来了革命性的进步。MOFs 的孔隙性结构使其能够高效地负载各种药物，从小分子药物到大分子生物制剂。

研究显示，特定的 MOFs 结构可以显著提高甲氨蝶呤等药物的载药量和释药效率，从而增强其抗风湿活性同时降低其毒性。MOFs 的表面可进行多种化学修饰，能够响应生物体内的特定生化信号，如 pH 变化或酶的存在，从而实现靶向释放。例如，某些 MOFs 可设计为在酸性肿瘤微环境中解离，释放载药，精准攻击癌细胞，减少对健康细胞的影响。MOFs 的生物相容性和生物降解性也是其在药物递送中广受关注的重要因素。通过选用合适的金属离子和有机配体，可以制备出对人体友好且在体内能自然降解的 MOFs 材料，这对于临床应用至关重要。MOFs 在药物递送中的应用包括抗癌药物的精确递送、抗炎药物的缓释和控制，以及生物成像和疾病诊断中的使用。

6.1 负载抗菌药物

耐多药细菌的出现一直被认为是一场影响全球人类的危机[1]。由于传统抗生素的局限性，迫切需要新的抗感染策略[2]。然而，随着临床需求与抗菌治疗创新之间的差距不断扩大以及膜通透性的障碍，难以对抗的革兰氏阴性菌，这些问题严重限制了抗菌策略的重组。

金属有机框架（MOFs）具有可调节孔径、高载药率、可定制结构和优越的生物相容性等优点。而且，MOFs 中的金属元素通常具有杀菌作用，使其在抗菌治疗应用中作为药物载体有独特优势。载药 MOFs 随着金属离子和有机配体的降解释放活性成分而表现抗菌活性。在该应用中，MOFs 粒子的形状和大小、金属活性中心的存在等机制都是不可忽视的重要因素。MOFs 材料的抗菌性能与其结构的坍塌密切相关。同时，MOFs 组分的释放速率取决于其结构稳定性，进而决定了其抗菌效果。例如，根据软硬酸碱理论[3]，当软酸与硬碱相互作用时，MOFs结构不那么稳定，这种组合更有可能导致结构崩溃和成分的释放。不同的材料由于其配位键而具有不同的稳定性，这影响了它们在体内的抗菌作用。

抗生素耐药性（AMR）是一场日益严重的全球危机，由于对现有药物的耐药性越来越大，无法治愈或极难治疗的细菌感染越来越多。据预测，到 2050 年 AMR 将成为导致死亡的主要原因。除了在预防策略和感染控制方面的持续努力外，还在持续研究开发新型疫苗、抗菌剂和优化诊断方法以解决 AMR。然而，开发新的治疗剂和药物可能是一个漫长的过程。因此，世界范围内正在进行并行的努力，以开发用于优化药物输送的材料，以提高疗效并最大限度地减少 AMR。由于此类材料能够破坏病原体的许多基本成分，如表面功能化使其能够自我消毒或防污，因此纳米粒子的开发具有良好的抗菌特性。

MOFs 由于其结构特点，能够稳定、可持续地释放金属离子，因此被认为是良好金属离子供体。在选择 MOFs 材料的金属颗粒时，应首先考虑具有良好抗菌活性的金属离子，但在运输过程中和降解后的毒性是另一个需要考虑的问题。由于许多金属对生命的生物化学和新陈代谢都是不可或缺的。因此，金属离子至少应该具有以下两个特性：抗菌活性和最小毒性。然而，当这些必要的金属过量存在时，它们对细胞造成的伤害可能是致命的。目前发现，在人体中发现的浓度最高的微量元素是铁、锌和铜[4]，它们参与如催化和离子传递等重要的细胞代谢活动。锌[5, 6]、铜[7, 8]和银[9]由于其强大的抗菌性能和低毒性，常用于生物医学领域。

基于 MOF 的抗菌纳米平台通常包含以下一个或多个特征：

（1）MOFs 中的金属节点（如 Ag、Cu、Ni、Zn 等）表现出抗菌活性，可以进行缓释降解过程中实现高效的抗菌效果。

（2）优秀的遏制抗菌分子基于高表面积和可调孔径。

（3）用于构建 MOFs 框架的有机配体，如卟啉衍生物，提供可能结合抗菌分子位点。

6.1.1 MOFs 本身作为抗菌剂

传统的基于抗生素的细菌感染治疗仍然是医学上最困难的挑战之一，因为滥用会导致多重耐药性的威胁。为了解决这些问题，有必要开发一种有效的抗菌剂，可以小剂量使用，同时最大限度地减少多重耐药性的发生。MOFs 与传统抗生素不同，可以通过释放金属离子具有很强的抗菌活性。MOFs 中的金属节点本身就具有显著的抗菌效果，如银（Ag）、铜（Cu）、锌（Zn）等。这些金属离子能够干扰微生物的细胞功能，包括破坏细胞膜的完整性、与细胞内酶反应，或干扰微生物的 DNA 合成。由于 MOFs 结构的特殊性，它们能够稳定并持续地释放金属离子，这种缓释性质使 MOFs 在长时间内保持抗菌活性，从而有效控制和减少细菌生长。

Zhang 成功合成了 CuZn-MOF-74，并评估了含有聚甲基丙烯酸丁酯（PBMA）和 CuZn-MOF-74 的 UV 固化 PBMA 涂层在铜离子稳定释放和防污性能方面的功效[10]。由于金属离子与有机配体之间的配位以及 CuZn-MOF-74 中 Cu 和 Zn 阳离子之间的相互作用，PBMA/CuZn-MOF-74 涂层表现出稳定的铜离子释放性能。历时 21 天的抗菌实验结果表明，MOF-74 的添加提高了涂层的抗菌性能，且抗菌率可达 98%以上。涂层的抗菌机制主要包括开放金属位点的毒性作用、释放铜离子以及 ROS 增加引起的氧化应激。

Kim 等开发了一种光活性 MOFs 衍生的钴-银双金属纳米复合材料（Ag@CoMOF），其方法是通过纳米级电流置换将银纳米粒子简单地沉积在钴基 MOFs 上[11]。纳米复合结构在水相中不断释放抗菌金属离子（即 Ag 和 Co 离子），并表现出 Ag 纳米粒子的强光热转换效应，同时在近红外（NIR）光下温度快速升高 25～80℃。与常用的化学抗生素相比，使用这种基于 MOFs 的双金属纳米复合材料，抗菌活性提高了 22.1 倍，大肠杆菌和枯草芽孢杆菌在液体培养环境中对细菌生长的抑制作用增强了 18.3 倍。

Lelouche 和他的同事提出了两种基于抗菌三羧酸盐连接体和锌或铜阳离子新型 MOFs——具有抗菌特性的 IEF-23 和 IEF-24。评估它们对表皮葡萄球菌和大肠杆菌的抗菌活性。这些细菌是传播最广泛的病原体之一，更容易产生抗菌药物耐药性。MOFs 对两种菌株均具有活性，且对表皮葡萄球菌表现出更高的活性[12]。

MOFs 的高表面积和可调节的孔径使其能够有效载入和缓慢释放抗菌剂。这些特性使 MOFs 能够提高药物的局部浓度并延长作用时间成为潜在的抗菌药物载体。MOFs 不仅可以作为纯金属离子的来源，还可以与其他抗菌分子一起使用，如抗生素或天然抗菌剂。这种结合可以提高抗菌效率，并可能对多重耐药细菌株显示更强的抗菌活性。据报道，Yang 等设计了一种含石墨状碳结构的银掺杂 MOFs 衍生物（C-Zn/Ag），该衍生物具有广谱光吸收和高效的光热转换能力[13]。在辐照下近红外 C-Zn/Ag 的抑菌率为 76%，而仅释放锌离子的 C-ZIF 的抑菌率为 43%，实验结果表明协同作用提高了抗菌性能。

以上研究表明 MOFs 不仅在药物递送系统中展现出优异的性能，也在直接作为抗菌剂方面展现了潜力。这些特性支持 MOFs 在抗菌领域的应用，特别是在应对抗生素耐药性日益严重的今天，MOFs 提供了一种新的策略来增强传统抗菌治疗的效果。

2011 年，Jaffres 和同事报道了一种通过持续释放 Ag^+有效杀灭细菌[14]，它由 Ag^+和含有羧酸和膦酸官能团的 3-磷酸苯能酸组成。Ag-MOF 对人类红细胞的毒性可以忽略不计。抗菌结果显示，Ag-MOF 对六种类型的细菌显示广泛的杀菌效果，包括三个金黄色葡萄球菌菌株 RN4220，Newman 和耐甲氧西林金黄色葡萄球菌（MRSA），一个大肠杆菌菌株 MG1655，两个铜绿假单胞菌菌株 PA130709 和 PA240709，且采用阴极剥离伏安法测定了 Ag^+缓释后具有抗菌活性。

Zamaro 及其团队研究发现，由 Cu^{2+}和三聚体酸（TMA）配体组合而成的 Cu-MOF，即 HKUST-1，显示出对酿酒酵母和假念珠菌的显著抑制效果。这种抑制作用与 Ag-MOF 的作用机制相似，主要归因于 MOFs 框架的降解过程中 Cu^{2+}的逐渐释放，进而形成表面外的 Cu 框架。Cu-MOF 通过这种方式实现了金属离子的缓释，为 MOFs 在抗菌治疗领域的应用开辟了新的可能性和前景[15]。

Rauf 等通过微波辅助水热合成了宏观纳米纤维铜基配位聚合物[Cu(HBTC)(H_2O)$_3$]。通过对大肠杆菌和金黄色葡萄球菌的菌落计数测定来测定其抗菌活性。在相同浓度下，纳米纤维的抑制率分别为 99.9%和 99.1%，而单独使用铜的抑制率为 24%。配体 H_3BTC 对细菌没有明显的抑制作用，这表明铜基配位聚合物[Cu(HBTC)(H_2O)$_3$]释放出铜离子，便于灭菌[16]。Li 等还报告了制备银纳米颗粒（Ag NP）装饰的二维卟啉 MOFs 纳米片（表示为 Ag/Co-TCPP NSs）的高效联合抗菌治疗。在 660 nm 激光照射下，Ag/Co-TCPP NSs 局部产生单线态氧（$^1O_2$），并促进 Ag NPs 部分降解以持续释放有毒的 Ag^+，达到协同杀菌效果（图 6.1）[17]。

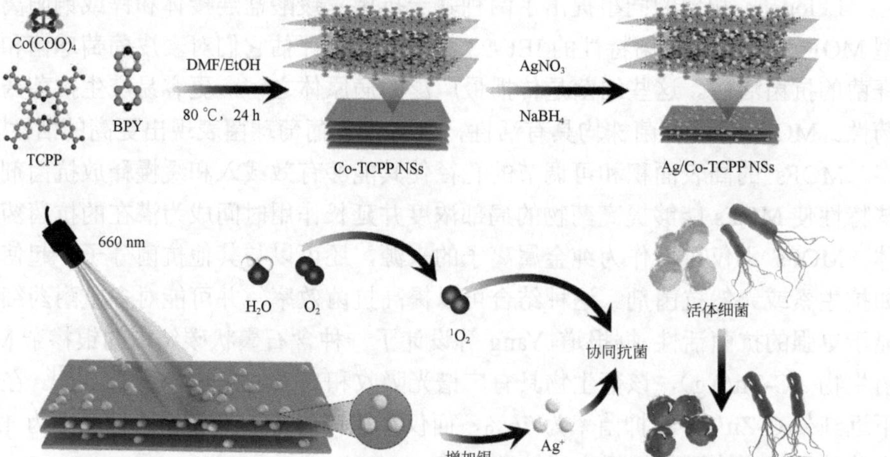

图 6.1　Ag NPs 协同杀菌[17]

6.1.2　MOFs 作为抗菌材料载体

2017 年，Wu 和同事报道了一种 MOF-53（Fe）@Van，用于持续地抗菌治疗无毒的 MOF-53（Fe）系统[18]。MOF-53（Fe）载体在酸性条件下（pH 值分别为 6.5 和 5.5）的降解作用可以忽略不计。实验结果表明，MOF-53（Fe）@Van 具有可控的药物释放行为、显著的抗菌效率（高达 90%）和良好的化学稳定性以及细胞毒性。

Sava Gallis 等报道了另一种缓释治疗平台，即将抗菌药物头孢他啶通过简单的封装策略成功地装入 ZIF-8 中，用于持续治疗细胞内感染[19]，并显示出药物的长期释放长达 1 周。头孢他啶@ZIF-8 作为治疗平台对 A549 细胞和 RAW 264.7 细胞表现出良好的生物相容性，对大肠杆菌具有良好的抑菌效果。更为重要的是，ZIF-8 由于内在性发射特性，通过共聚焦显微镜三维重建叠加，明确地证明了 ZIF-8 的直接内化和细胞内的药物释放。

Huang 小组研究提出了一种将羧甲基壳聚糖（CMCS）与 HKUST-1 交织的新策略，构建了含富马酸二甲酯的 HKUST-1@CMCS，具有生态友好、可回收、长效、智能的优越的长效抗菌活性。此外，他们还研究开发了一种新的磁性框架 Fe_3O_4@PAA@ZIF-8 用于传递环丙沙星（CIP）抗生素[20]。Cai 等成功以 MOFs 作为载体，负载了天然抗菌剂姜黄素（Curcumin，Cur），制得了载药 MOF（Cur@ZIF-8）。随后，他们采用浸涂法将 Cur@ZIF-8 修饰于真丝缝合线表面，进而开发出具备抗菌与促愈功能的真丝缝合线。实验结果显示，经过这种抗菌处理的缝合线能够有效阻止微生物在其表面的黏附，从而显著降低了手术部位感染的发生率，为医疗缝合提供了更为安全、有效的选择[21]。

诺氟沙星（Nfx）根据生物药剂学分类系统（BCS）为Ⅳ类药物，是一种重要的抗菌氟喹诺酮类药物，存在溶解度和渗透性较低的问题。这种较差的理化性质的药物分子导致递送不良，是制药行业临床开发的重要关注点。Yadav 提出了一种通过将药物分子负载到生物相容性金属有机框架（MOFs）MIL-100(Fe)的多孔平台上来传递 NFX 的概念性新方法[22]。通过用 PEG 包覆载药 MIL-100(Fe)来进一步控制诺氟沙星的释放 PEG（Nfx@MIL-100(Fe)）。通过毒性研究测试了两种药物递送系统（DDS）NFX@MIL-100(Fe)和 PEG（NFX@MIL-100(Fe)）的生物相容性。

Bhat 等报告了一种新的左氧氟沙星与 MOFs 结合为 Levo@Zn-NMOF[23]。与纯左氧氟沙星相比，与之偶联的 Zn-NMOF（Levo@Zn-NMOF）对革兰氏阳性和革兰氏阴性细菌的抗菌效果增强。

Shakya 等以环糊精金属有机框架（CD-MOFs）为载体与溶解的磺胺嘧啶（SD）共同递送超细纳米银以此表现出优于不溶性磺胺嘧啶银的抗菌效果。CD-MOFs 中的丰富羟基部分用于将 Ag 前体还原为 4～5 nm 的银纳米颗粒（Ag NPs），并固定在纳米腔内。微孔 CD-MOFs 有助于将 SD 分子包含在γ-环糊精（γ-CD）分子对的疏水腔中，亲水性 CD-MOFs 可以很容易地溶解在伤口区域的渗出液中，从而释放出药物。这种方法将 SD 的水溶性提高了 50 倍，并增强了 SD 的释放及其抗菌活性。CD 框架防止了纳米银颗粒的聚集并稳定了粒径并增强了治疗效果[24]。

Asadollahi 开发了海藻酸钠-玉米醇溶蛋白（SA-ZN）纳米复合材料修饰的金

属有机框架，并将其用作柳氮磺吡啶（SSZ）递送的有效纳米载体，减少了 SSZ 的副作用，从而提高了其生物利用度。还研究了样品对金黄色葡萄球菌和铜绿假单胞菌的抗菌特性。结果表明，与 SSZ 相比，载药 MOF/SA-ZN 对金黄色葡萄球菌革兰氏阳性菌具有更高的抗菌活性[25]。

Huang 等构建了一种用于治疗 MRSA 感染的新型纳米药物递送系统，并评估该系统的治疗效果和生物毒性。万古霉素的包封率（EE）和装载效率（LE）分别为 81.0%和 64.7%。PLT@Ag-MOF-Vanc 是一种具有良好生物相容性的 pH 响应纳米药物递送系统。PLT@Ag-MOF-Vanc 在体外对常见临床菌株表现出比游离万古霉素更好的抗菌活性[26]。

Wei 等提出了一种可提供适用于治疗细菌感染的抗生素的γ-环糊精金属有机框架。γ-环糊精金属有机框架是利用 γ-环糊精和钾离子通过超声方法开发的。负载抗生素的γ-CD-MOFs 在体外和体内均对哺乳动物细胞和组织显示出无毒和完美的生物相容性[27]。He 等[28]还构建了以改善纳米银（Ag NPs）易沉淀和高细胞毒性等抗菌缺陷的一种基于沸石咪唑酯框架（ZIF-8）的新兴药物递送系统。在弱酸性细菌微环境（pH=6.4）下，AgNPs@ZIF-8 表现出更优异的抗菌效果。

6.1.3　光动力学、超声动力学抗菌

抗菌光动力学治疗（APDT）是一种很有前途的替代抗感染疗法，其基于细菌和光敏剂进行光敏治疗。APDT 的机制如下：光敏剂暴露于共振波长光后产生活性氧（ROS），具有很高的杀死细菌的潜力。和抗生素相比，ROS 可以作用于多个细菌目标，提供更多的机会杀死细菌，该策略可诱导耐药细菌。

Zhang 和他的同事发现了一种水溶性环糊精（CD-MOFs）作为合成模板 GS5-CL-Ag@CD-MOF。该模板带有 Gly-Arg-Gly-Asp-Ser（GRGDS）肽功能化，载有超细的 Ag NPs 以促进抗菌作用和伤口愈合。CD-MOFs 的孔径为 0.78 nm，以γ-CDs 为配体和 K^+组成，并且固定在 CD-MOFs 孔内的 Ag NPs 具有 2 nm 的超细粒径、高水分散度，对革兰氏阴性和革兰氏阳性细菌均有有效的抗菌作用。伤口愈合结果表明，基于 MOFs 的混合材料具有增强的止血效果和有效的杀菌能力，为合理设计有效的伤口修复装置显示出了良好的潜力[29]。

由于抗生素耐药性降低了感染伤口愈合的有效性，因此有必要制定一种新的策略来促进感染伤口愈合从而不使用抗生素。Zhang 和他的同事开发了一种基于海藻酸盐和聚丙二醇的锌金属有机框架（MOFs）热敏水凝胶，可增强抗菌作用并促进感染伤口愈合。更重要的是，通过将 Zn-MOFs 与光动力学治疗（PDT）相结合的新策略，其可减少炎症并促进胶原蛋白沉积和再上皮化，从而促进感染伤口的愈合[30]。

用抗菌纳米材料对电纺垫进行功能化是防止细菌定植并加速感染伤口愈合的

一种有吸引力的策略。与传统的依靠杀菌剂的抗菌策略相比，光动力学治疗（PDT）由于其高效和可控性而受到了广泛的关注。Tang 等构建了卟啉金属有机框架（MOFs）纳米复合材料（PS）放在 HKUST-1 变型的笼子中，可有效产生单重态氧（$^1O_2$）在光照射下灭活植物病原体。结果表明，所制备的 PS @ MOFs 具有约 12%（质量分数）的 PS 负载率，并且在体外对三种植物病原性真菌和两种病原性细菌具有优异的广谱光动力学抗菌活性[31]。

超声动力学治疗（SDT）利用超声（US）激活超声增敏剂，产生高细胞毒性的反应性氧化物（ROS），这在根除深层细菌感染方面取得了巨大的成功。然而，声增敏剂在生物膜中的有限渗透和低扩散效果严重损害了 SDT 的治疗效果。有研究设计了一种近红外（NIR）光驱动纳米马达（MOF@AuDNase I），具有较高的电子−空穴对分离效率，用于对抗金黄色葡萄球菌（*S. aureus*）生物膜。他们将金纳米颗粒（Au NPs）沉积在球形金属有机框架（MOFs）上，然后将脱氧核糖核酸酶（DNase I）共价固定在 MOF@Au 表面。在 NIR 激光照射下，MOF@Au NPs 表现出有效的主动运动，并在 15 min 内快速穿透深层生物膜[32]。超声介导的药物输送技术的最新进展表明可以改善药物和基因在空间上受限的向靶标组织的输送，同时降低全身剂量和毒性。

提供协同光热/光动力学治疗（PTT/PDT）的抗菌表面为对抗细菌感染提供了有希望的途径，但这通常需要耗时且烦琐的化学制备程序。在此，Gao 等通过 ZIF-8 颗粒和多巴胺（DA）之间惊人的快速且可持续的螯合诱导自组装，实现了具有 PTT/PDT 抗菌性能的金属有机框架（MOFs）复合膜[33]。与传统的 DA 聚合不同，通常在反应时间较长的碱性介质中（例如超过 20 小时），在中性（pH=7）条件下，可以在 0.5 小时内在各种散装材料上快速组装以 DA 作为接头的坚固且通用的 MOFs 复合膜，这种 PTT/PDT 抗菌膜还表现出优异的体外生物相容性。

Qian 等将抗菌剂 RB 一步封装到沸石咪唑酯框架（ZIF-8）中以获得光动力学抗菌性 RB @ ZIF-8 纳米粒子，然后与 PCL 基质共混以通过共电纺丝制备复合聚合物纳米纤维。通过调节 PCL 中 RB @ ZIF-8 的含量，在纳米纤维表面上存在足够的 MOF 颗粒。得益于可见光照射后产生的活性氧（ROS），纳米纤维膜能够抑制金黄色葡萄球菌和大肠杆菌[34]。

锆基 MOFs 颗粒 UiO-66 在可见光下产生适量的活性氧，对促进 DA 聚合起着重要作用，从而大大促进了 MOF 膜的形成。Hao 等通过将玫瑰孟加拉引入 MOFs 膜[35]，实现了对革兰氏阳性金黄色葡萄球菌和革兰氏阴性大肠杆菌具有出色的光动力学抗菌活性。据报道，Yang 等设计了一种含石墨状碳结构的银掺杂 MOFs 衍生物（C-Zn/Ag），该衍生物具有广谱光吸收和高效的光热转换能力。近红外 C-Zn/Ag 在辐照下的抑菌率为 76%，而仅释放锌离子的 C-ZIF 的抑菌率为 43%，表明协同作用提高了抗菌性能[13]。一些研究人员通过将抗生素封装在光敏剂修饰的 MOFs（Van@ZIF-8@PDA）中，开发了一种 NIR/pH 双刺激响应性抗菌制剂。聚多巴胺（PDA）在近红外光下具有热效应，具有较高的抗菌作用，可以与 ZIF-8 的降解协同释放锌离子和负载药物[36]。MOFs 材料抗菌研究策略多种多样，而且还会和其他功能联合起效（表 6.1）。

表 6.1　MOFs 抗菌应用实例

MOF	抗菌药物	中心离子	结果	参考文献
HKUST-1@CMCS	羧甲基壳聚糖（CMCS），富马酸二甲酯	Cu^{2+}	HKUST-1@CMCS 非常智能和长效（384 h，0.04 mol/L PBS），HKUST-1@CMCS 可以二次利用，循环使用的 HKSUT-1@CMCS 在一个使用周期内仍然保持着智能响应特性	[20]
Phy@ZIF-8	Physcion（Phy），化学名称为 1,8-二羟基-3-甲氧基-6-甲基-9,10-蒽二酮，在何首乌中提取	Zn^{2+}	在 pH 为 5.0 时，ZIF-8 负载 Phy 的释放量为 88.72%，约是 pH 为 7.4 时生理系统（27.61%）的 3 倍	[37]
壳聚糖包裹 MIL-53	万古霉素	Fe^{3+}	通过壳聚糖包覆的铁 MOFs 传递来增加万古霉素对金黄色葡萄球菌耐药菌株的杀菌潜力	[38]
Tebuc@ PCN@P@C	丁康唑	Zr^{4+}	Tebuc@PCN@P@C 微胶囊具有 PDT 和抑制微生物剂双重活性	[39]
Fe_3O_4@PAA@ZIF-8	环丙沙星（Cip）	Fe^{3+}，Zn^{2+}	构建的框架在机体生理条件下具有运输和药物释放的能力	[40]
Bio-MOFs	萘啶酸	Mg^{2+}，Mn^{2+}	合成的 MOFs 显示出足够的细胞毒性和较高的抗菌活性，特别是对革兰氏阴性菌	[41]
水凝胶@Cu-MOF 1		Cu^{2+}	水凝胶@Cu-MOF 1 没有细胞毒性作用，但在最低杀菌浓度下具有 99.9%的抗菌作用	[42]
水凝胶@Co-MOF 2		Co^{2+}		
Zn-MOF	甲基丙烯酸透明质酸（MeHA）	Zn^{2+}	Zn-MOFs 释放的锌离子对细菌胶囊和氧化应激的损伤能力，该 MNs 阵列具有良好的抗菌活性，以及相当强的生物相容性	[43]
wool-MOF-SO 和 CO	Salvia Officinalis （SO）和 Calendula Officinalis（CO）	Zr^{2+}	将丹参（SO）和金盏花（CO）提取物装载到改性织物上，其吸收能力分别提高了 1154%和 1842%	[44]
$ZnFe_2O_4$/IRMOF-3/GO/四环素	四环素	Zn^{2+}	对金黄色葡萄球菌和铜绿假单胞菌具有比单独使用更好的抑菌活性	[45]

6.2　抗炎药物和疫苗的递送

6.2.1　装载抗炎药物

甲氨蝶呤（MTX）已被用作治疗类风湿性关节炎（RA）的锚定药物，而慢性使用甲氨蝶呤的患者则存在严重的副作用[46]。为此，通过纳米药物靶向传递甲氨蝶呤已经引起了人们的极大兴趣。有研究将 MTX 与单宁酸（TA）按 2∶1 的比例偶联，然后与铁离子（Fe^{3+}）配位从而对透明质酸（HA）进行表面修饰，得到的 MOFs 达到超高的载药量（45%）和持续的释药[47]，可选择性识别病变细胞的抗炎作用。体内治疗评价表明，MOFs 不仅可通过增强甲氨蝶呤的抗风湿活性，同时还通过靶向给药来降低其毒性作用，从而提高治疗指标。

Li 等[48]以 UiO-66 金属有机框架（MOFs）作为酮洛芬传递系统，创新地应用于骨关节炎（OA）的治疗中。在研究中，他们巧妙地将两种不同类型的 NH_2 和 NO_2 官能基团引入 UiO-66 框架中，以实现对药物的高效负载与可控释放。通过软骨细胞毒性实验，他们验证了所合成的 MOFs 载体具备良好的生物安全性，确保了其作为药物传递载体的可靠性与有效性。这一研究为骨关节炎治疗提供了新的思路和方法，展现了 MOFs 在医药领域的广阔应用前景。

目前以碱土离子为中心离子的 MOFs，仍有待开发。在骨关节炎治疗中，Sr 在减少软骨退行性变和减少软骨细胞凋亡方面发挥着重要作用。Li 等设计开发了 Sr/HCOOH-MOF[49]，载药和释放实验证明 Sr/HCOOH-MOF 上只负载 3%的酮洛芬，8 h 后再释放约 80%的酮洛芬。Sr/HCOOH-MOF 在给定剂量内具有生物相容性，并可显著抑制 IL-1β、iNOS 和 RANKL 的基因表达。

有研究通过单宁酸（TA）和 Fe^{3+}之间的配位制备抗 TNF-α siRNA[50]，通过简单的超声处理加载，实现了与阳离子载体相当的高加载能力。MOFs 通过质子海绵效应实现 siRNA 的快速内/溶酶体逃逸从而表现出优异的生物相容性，可以有效下调细胞因子。这种纳米药物可以消除广谱的活性氧，进而与 siRNA 联合使用以增强 RA（类风湿关节炎）治疗将 M1 巨噬细胞重新极化为抗炎 M2 表型。MOF 进一步用牛血清白蛋白（BSA）修饰，可以更好地向 RA 关节和患病巨噬细胞靶向递送。

治疗剂的软骨靶向递送仍然是骨关节炎（OA）治疗的有效策略。最近，越来越多的报道称清除活性氧（ROS）和激活自噬可有效治疗 OA。Xu 等[51]首次设计了一种基于金属有机框架（MOFs）修饰的介孔多巴胺（MPDA）的双药物递送系统，该系统由负载于中孔的 MOFs 外壳上的雷帕霉素（Rap）和胆红素（Br）组成。结果表明，可以通过近红外（NIR）激光刺激从 RB@MPMW 中连续释放两种药剂。

Taherzade 等提出了基于水稳定且生物安全的羧酸铁 MOFs（MIL-100 和 MIL-127）、生物聚合物聚乙烯醇（PVA）和两种用于皮肤病的共封装药物[壬二酸（AzA）和抗生素和烟酰胺（Nic）作为抗炎剂]，用以开发先进的皮肤联合疗法[52]。该方法达到了优异的 MOFs 药物含量（MIL-100 和 MIL-127 的总量分别为 77.4%和 48.1%，质量分数），同时两种药物在 24 小时后几乎完全释放，且适合皮肤递送。制备的皮肤 PVA-MOF 制剂具有生物相容性，并保持高载药能力（MIL-100 和 MIL-127 的总药物含量分别为 38.8%和 24.2%，质量分数）。

MOFs 在生物医学中的应用仍然存在不足，例如有限的化学和胶体稳定性以及毒性。Zhao 等[53]报告了通过一锅法合成的结合纳米 MOFs（即 MIL-100（Fe））和超小超顺磁性氧化铁（USPIO）纳米粒子。两种纳米粒子的物理化学和功能特性的协同耦合赋予这些纳米物体有价值的特性，例如高胶体稳定性、高生物降解性、低毒性、高载药量以及刺激响应药物释放和超顺磁性。这种双峰 MIL-100（Fe）/磁赤铁矿纳米载体装载了抗肿瘤和抗炎药物（阿霉素和甲氨蝶呤）后，显示出很高的抗炎和抗肿瘤活性。

6.2.2 疫苗递送

疫苗接种在预防和治疗感染方面是非常有效的。与传统的灭活疫苗和减毒疫苗相比，基于蛋白抗原的疫苗具有更好的安全性且成本更低，但蛋白抗原容易降解，单独给药时免疫原较差。因此，基于抗原疫苗在生物医学中的应用受到了限制，必须使用输送载体和佐剂来改善免疫应答。弗氏完全佐剂（FCA）是最常见的佐剂，由于其对细胞和体液免疫具有特殊的强刺激作用，已广泛应用于动物疫苗中[54]。然而，FCA 可引起持续疼痛、局部组织坏死和注射部位的肿瘤样增生等严重的不良反应。目前，具有高安全性的明矾是 FDA 批准的用于人类疫苗的唯一佐剂。然而，明矾也有一些缺点，如细胞免疫能力不足和免疫原性较差。因此，迫切需要开发出效力较强但毒性较低的新型疫苗佐剂，提高疫苗的安全性和有效性。

Li 等[55]用 span-85 对 *c*-CD-MOF 进行了修饰，制备了 SP-*c*-CD-MOF 作为动物疫苗佐剂。将卵清蛋白（OVA）作为模型抗原包裹成颗粒来研究免疫反应（图 6.2）。SP-*c*-CD-MOF 在体内外均具有良好的生物相容性。免疫后的含 OVA 的 SP-*c*-CD-MOF 可诱导高抗原特异性 IgG 滴度和细胞因子分泌。同时，SP-*c*-CD-MOF 也能显著促进脾脏细胞的增殖，并激活和成熟了骨髓树突状细胞（BMDCs）。该研究显示了 SP-*c*-CDMOF 在疫苗佐剂中的应用潜力，并为疫苗佐剂的开发提供了新的思路。

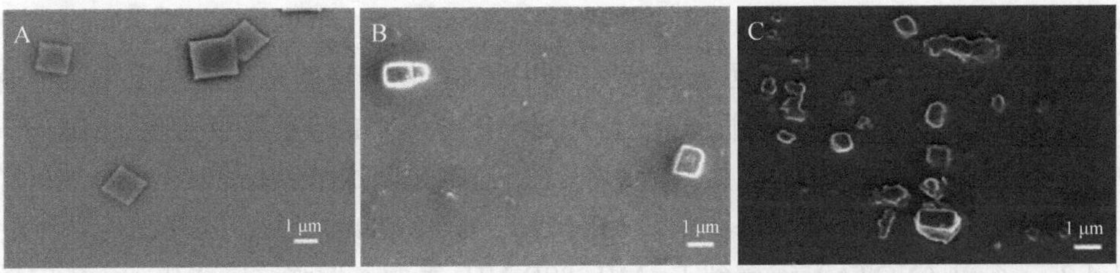

图 6.2　抗原包裹 CD-MOFs[55]

口服抗原接种疫苗面临着许多挑战，包括胃肠道（GI）蛋白水解和黏膜屏障。为了限制胃肠道蛋白的水解。Zheng 等[56]研究开发了一种生物模拟矿化铝基金属有机框架（Al-MOF）系统，该系统耐环境温度和 pH，可以协同作为传递载体，并在模型抗原卵清蛋白（OVA）上合成佐剂作为载体。有报道一锅方法通过异丙醇铝与 2-氨基对苯二甲酸配位反应合成纳米级 MOFs，以封装模型蛋白抗原卵清蛋白（OVA）[57]。该方法模拟了生物体分泌无机矿物质以形成外骨骼。OVA 周围铝基 MOFs 的仿生矿化（OVA@Al-MOFs）形成一个带正电荷的外骨骼笼，可能保护被包裹的抗原免受环境温度和 pH 以及胃肠道中高度降解环境的影响[58]。

6.3　慢性疾病治疗及肺部给药

MOFs 不仅用于药物载体，还能结合成像标记或靶向配体，因此能够被设计为多功能平台。例如，Fe-MIL-88B-NH_2 能够封装亚甲蓝并通过磁共振成像进行追踪，同时附着靶向剂增强对特定病理特征的靶向性。MOFs 的设计可以特别针对阿尔茨海默病等神经退行性疾病，通过卟啉配体等促进抗氧化作用和高效清除神经毒性金属离子，提高疗效。通过精确控制药物释放，可以使药物释放与生理需求同步，通过肺部直接给药，提高药物的局部浓度，减少全身副作用，尤其适用于肺部疾病和某些全身性疾病的治疗。此外，MOFs 可以被加工成适合吸入的粒子，具有良好的空气动力学特性，能够有效地到达深肺部位，增强治疗效果。这些特点使得 MOFs 在现代药物输送系统中很重要。特别是在面对需要精确控制药物释放和靶向治疗的慢性疾病管理中表现出独特的优势。

阿尔茨海默病（AD）[59]是一种神经退行性疾病，其主要症状表现为记忆力逐渐减退，同时伴随着感觉、思维、判断以及运动能力的明显受损。这一疾病严重影响了患者的日常生活和社交能力，为他们的家庭和社会带来了沉重的负担。尽管有了大量的研究，但 AD 发病的复杂分子机制仍不清楚。世界各地几乎所有的工业和研究机构所开发 AD 药物的尝试都以失败告终。目前的治疗方法只是缓解 AD 的症状，而不能阻止或治愈 AD 的进展[60]。因此，靶向治疗方法和其他策略可能更适合对抗复杂的疾病。

Zhao 等开发用于治疗阿尔茨海默病的靶向药物传递的多功能纳米尺度金属有机框架[60]。系统地选择磁性纳米材料 Fe-MIL-88BNH$_2$ 包封亚甲基蓝（MB，一种 tau 聚集抑制剂）并作为磁共振造影剂。随后，将靶向试剂 5-amino-3-（pyrrolo[2,3-*c*]pyridin-1-yl）isoquinoline（脱氟 MK6240，DMK6240）通过 1,4,7-三氮环甲烷-1,4,7-三乙酸（NOTA）连接到 Fe-MIL-88B-NH$_2$ 表面，增强过磷酸化 tau 靶向性及其表面特性使其具有出色的磁共振成像能力，并通过抑制过度磷酸化的 tau 蛋白聚集和阻碍神经元死亡，在体外和体内改善 AD 症状。

研究者们成功合成并筛选了四种以不同金属为节点的卟啉 MOF，旨在应用于靶向抑制 Aβ（amyloid-β）聚集的过程[61]。经过对活性氧（ROS）产生能力的评估，Hf-MOF 脱颖而出，被视为最佳的 Aβ氧化剂。为了揭示不同金属 MOFs 在产生 ROS 能力上的差异，研究者们采用了密度泛函理论进行了深入的计算和分析。在 MOFs 的结构中，卟啉配体扮演着多重角色。它不仅能够作为光氧化剂，有效抑制 Aβ的聚集，还能够作为铜螯合试剂，高选择性地移除神经毒性的铜离子，从而减轻对神经系统的损伤。此外，MOFs 的 Aβ富集能力使得光氧化效率得以最大化，同时降低了对正常细胞和组织的潜在损害。为了进一步增强 Hf-MOF 在细胞内环境中对 Aβ的靶向作用，研究者们巧妙地在其表面修饰了 Aβ靶向肽 LPFFD。这一改进不仅提高了光氧化效果，还增强了治疗的精准性。在 AD 线虫模型的实验中，Hf-MOF 展现出了显著的疗效。它能够显著减少 Aβ 诱导的细胞毒性，并有效延长线虫的生命周期。这一成果为 MOFs 在 AD 治疗领域的应用开辟了新的道路，展现了其作为多功能治疗试剂的巨大潜力。通过促进 MOFs 整合神经毒性的金属离子，提高 Aβ 的光氧化程度，有望为缓解 AD 症状提供新的策略和方法。

Zhao 等[62]研究设计了一种基于 MOFs 平台的独特的 MRI 和靶向给药系统。该表面修饰的 DMK6240 以结合过度磷酸化的 tau 蛋白。MB 被包裹在 Fe-MIL88B-NH$_2$ 的孔中，以抑制 tau 蛋白的聚集。由于 Fe-MIL88B-NH$_2$ 是 MRI 造影剂的理想选择，因此可以通过 MRI 监测其在病变区域的保留情况，可作为一种有效的治疗 AD 的重复靶向纳米载体。

酚类天然产物木兰醇通过 PC-12 细胞中的β-淀粉样蛋白毒性表现出神经保护特性，并在 TgCRND8 转基因小鼠模型中对认知缺陷有改善作用。Santos 等[63]研究评估了木兰醇和含木兰醇的 UiO-66（Zr）（Mag@UiO-66（Zr））对β分泌酶和 AlCl$_3$ 诱导的神经毒性的抑制作用。与木酚相比，含木酚的 DDS UiO-66（Zr）除了损伤逆转外，对中性粒细胞浸润和凋亡神经元也表现出增强的神经保护活性。因此，MOFs 是生物可利用性差的药物的很有前途的药物传递平台。

Yang 等研究开发了一种多酶金属有机框架（MOFs），该框架结合了胰岛素和葡萄糖氧化酶，并通过钴掺杂的 ZIF-8[64]进行负载。这一创新设计旨在实现无痛葡萄糖介导的透皮给药，通过刺激响应微针（MNs）实现药物的精准传递。这一方法有望为糖尿病患者提供一种更为便捷和舒适的治疗方式，具有重要的临床应用价值。游离的 Co^{2+}离子将被 EDTA-SiO_2 纳米颗粒螯合，并通过剥离 MNs 去除。所获得的模拟多酶基于 MOFs 的 MNs 对葡萄糖浓度有良好的依赖性，既不泄漏过氧化氢和 Co^{2+}离子，还有足够的硬度穿透皮肤。

糖尿病创面愈合是生物医学领域面临的主要挑战之一。常规单药治疗效果不理想，给药效果受到渗透深度的限制[65]。Yin 等[66]开发了一种基于镁有机框架的微针贴片（表示为 MN-MOF-GO-Ag），可以实现经皮传递和联合治疗糖尿病伤口愈合（图 6.3）。多功能有机镁框架（Mg-MOFs）与聚(γ-谷氨酸)（γ-PGA）水凝胶混合，加入 MN-MOF-GO-Ag 的尖端后 Mg 在真皮深层缓慢释放 Mg^{2+}和没食子酸，释放的 Mg^{2+}诱导细胞迁移和内皮小管形成。而没食子酸是一种活性氧清除剂，促进抗氧化。MN-MOF-GO-Ag 的背衬层由 γ-PGA 水凝胶和氧化石墨烯-银纳米复合材料（GO-Ag）制成，对加速伤口愈合具有良好的抗菌效果。MN-MOF-GO-Ag 对糖尿病小鼠模型的全层皮肤创面愈合有治疗作用，用 MN-MOF-GO-Ag 治疗的小鼠的伤口愈合有显著改善。

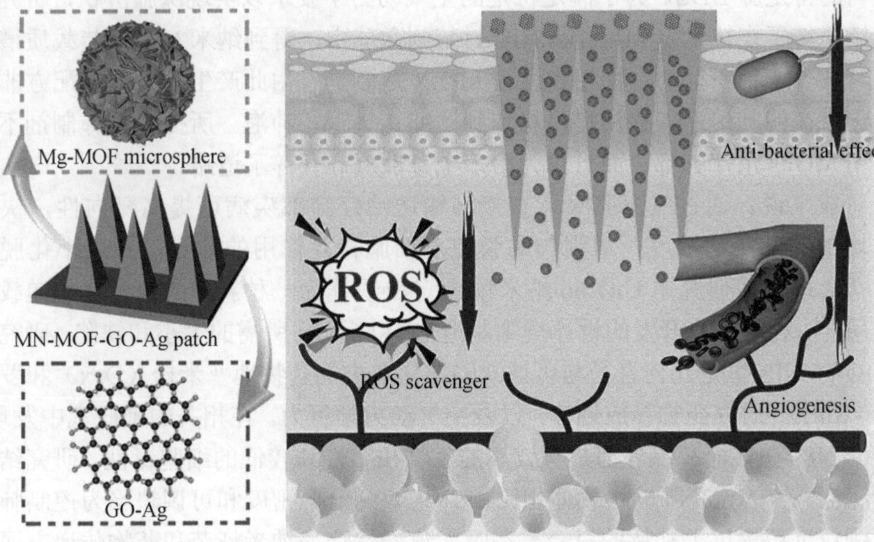

图 6.3　基于镁有机框架的微针贴片[66]

此外，靶向糖尿病病理相关信号通路的抑制剂对糖尿病伤口愈合具有巨大的治疗潜力。Sun 等研究人员将晚期糖基化终末产物（RAGE）抑制剂受体装入基于钴（Co）的 MOF（ZIF-67）中可制备得到 FZ@ZIF-67 纳米颗粒[67]。FZ@ZIF-67 NPs 可以以可控的方式双重传递超过 14 天。体外研究表明，FZ@ZIF-67 NPs 不仅通过传递 Co 离子增强血管生成，还促进 M2 巨噬细胞极化，并抑制高糖或炎症诱导的血管生成损伤。

肺是局部和全身传递活性成分（如药物、核酸和多肽）的一个非常有吸引力的目标。不仅治疗局部肺部疾病，而且治疗非肺部疾病。肺在呼吸系统疾病的治疗中提供了一个直接和精确的进入所需的位置，同时为药物的吸收提供了一个巨大的表面积和一个相对较低的酶控制环[68]。肺途径有几个优点：

（1）它是无创的，限制雾化局部效应治疗药物进入血液，从而减少与非肺给药相关的不良全身副作用[69]；

（2）肺有较大的表面（成人 70～140 m^2）以及极薄（0.1～0.2 μm）和高度血管化的肺泡-毛细血管膜，有利于药物在肺泡内吸收；

（3）避免了肝脏首通效应。

因此，在过去的几年中，新的递送设备（主要是喷雾器和干粉吸入器）[70]和胶体药物载体（如脂质体、纳米颗粒和树状大分子），被用于将活性成分递送入肺[71]。然而，它们的低产量、长期不稳定和随后的突发性药物释放以及肺渗透不足，这些是需要解决的主要缺点。

Fernández-Paz 等[72]报道了一种简单而有效的策略来开发一种基于 MIL-100 复合材料的特定肺 DDS。为了满足特定的空气动力学要求以实现深肺沉积，研究使用连续喷雾干燥技术将 MIL-100 NPs 封装到微球中，得到纳米 MOFs 装载质谱干粉，开发了一种新型的生物相容性肺 MOFs 基制剂。由此产生的微球形配方非常适合于潜在的肺治疗，该治疗可均匀到达细支气管和肺泡。所提出的肺制剂不仅为使用 MOFs 治疗肺部，而且也为治疗全身性疾病开辟了道路。

一般来说，通过气溶胶吸入直接肺输送治疗肺部疾病可提高靶向性，快速吸收增加局部药物浓度，增强与黏膜免疫细胞相互作用的能力，并最小化脱靶效应[73]。Jarai 等研究了 UiO-66 纳米颗粒（NPs）作为一种新型的肺药物传递载体的使用[74]。利用先前开发的设计规则制造 UiO-66 NPs 所需的大小和缺陷，研究合成 UiO-66 NPs 恒定几何直径与可调缺陷评估缺失的药物地塞米松（Dex）和罗丹明 B（RhB），研究药物释放规律，以及空气动力学行为。在相关的肺研究中发现，UiO-66 NPs 在体内和体外都具有较高的生物相容性和较低的细胞毒性。研究结果表明，与传统的无孔肺输送载体相比，UiO-66 的低容密度和可调缺陷为控制肺渗透和颗粒沉积提供了独特的机会，在吸入免疫治疗、纳米疫苗和药物传递方面有潜在的应用。通过对 UiO-66 用于肺给药的综合评估，证明了 UiO-66 NPs 作为一种新型气溶胶平台的可行性，具有潜在应用于广泛的呼吸系统疾病。

吸入给药可以克服与口服给药相关的水溶性差、吸收率低或生物利用度低等问题。在目前的研究中，Li 等研究使用了通过干粉吸入器实现靶向肺给药的可吸入颗粒大小的γ-环糊精金属有机框架（CD-MOFs）[75]，其具有立方形态和良好的空气动力学特性，选择活性天然化合物 paeonol（PAE）作为治疗急性肺损伤（ALI）

的模型药物。此外，姜黄素被装入环糊精基的金属有机框架（CD-MOFs）中用于肺输送[76]。与微粉姜黄素相比，姜黄素负载 CD-MOFs（Cur-CD-MOFs）具有优异的空气动力学性能，这是由于 CD-MOFs 独特的多孔结构和较低的密度。溶出度试验表明，Cur-CD-MOFs 的药物释放速率明显快于微粉化的姜黄素。全原子分子动力学模拟结果表明，姜黄素分子被载入 CD-MOFs 的疏水腔内或进入大的亲水腔内形成纳米团簇。Cur-CD-MOFs 提高的润湿性和姜黄素在 CD-MOFs 多孔内部的独特空间分布特征可能有利于提高溶解速率。而且 DPPH 自由基清除试验表明，Cur-CD-MOFs 具有显著的抗氧化活性。

6.4 负载核酸、蛋白质

生物大分子药物已成为治疗人类疾病的一类重要治疗剂。考虑到它们在人体内的高度降解倾向，选择合适的递送系统是确保生物大分子药物在体内治疗效果的关键。作为一类新兴的超分子“主体”材料，金属有机框架（MOFs）在孔径可调、包封效率、药物释放可控、表面功能化简单和良好的生物相容性方面表现出优势。因此，基于 MOFs 的主客体系统已被广泛开发为一类新的灵活而强大的平台。

金属有机框架（MOFs）的精确稳定性能够有效封装各种小分子药物和大分子药物，例如核酸和蛋白质。在用活性靶向部分进行表面修饰后，由于内表面积大，MOFs 可以特异性地将大量有效载荷转移到作用部位。MOFs 与生物大分子、病毒和细胞相结合已成为新型生物复合材料，可应用于药物输送、生物传感、生物样本保存以及细胞和病毒操纵。

6.4.1 负载核酸

核酸尚未被广泛认为是被药物递送的活性物质。实际上，未修饰的核酸在酶学上是不稳定的。对于细胞摄取和有效载荷封装而言亲水太强，并且可能引起意想不到的生物学反应，例如免疫系统激活和血液凝固途径的延长[77]。然而，近来，围绕核酸的三个主要发展领域使得值得重新考虑其在药物递送中的作用。这些领域包括 DNA、RNA 纳米技术[78]、多价核酸纳米结构[79]和核酸适体[80]，它们分别提供了以无与伦比的结构控制水平改造纳米结构的能力，完全颠倒了线型/环状核酸[77]的某些生物学特性，并使用全核酸构建体实现抗体水平的靶向。

核酸包括脱氧核糖核酸（DNA）和核糖核酸（RNA），其在遗传信息的存储和表达中起着重要的作用[81]。通常情况下，将核酸整合到 MOF 纳米载体中可以有效地保护其免受降解，同时还能加快其在细胞内的摄取速度[82]。此外，通过对 MOFs 纳米颗粒与核酸的表面进行修饰，可以创造出空间和静电上的聚集位阻，从而显著增强其胶体的稳定性。这一方法不仅提高了核酸的生物利用度，还有助于实现更为稳定且高效的核酸传递系统，为生物医药领域的应用提供了有力支持[83]。

MOFs 可以通过金属-磷酸盐配位相互作用，DNA 主链磷酸盐和 MOFs 上的不饱和锆位点之间的内在多价配位或封装来装载核酸。MOFs 提供高密度位点，其表面很容易被末端磷酸修饰的寡核苷酸功能化，可以很容易地与提供多功能的纳米颗粒结合[84]。

2014 年，Mirkin 和同事[85]报道了一种核酸-MOFs 偶联纳米颗粒的制备方法。他们首先通过溶剂热合成技术成功获得了叠氮化物功能化的 UiO-66-N_3 纳米颗粒。随后，利用无铜菌株促进的点击反应，将二苄基环辛烯（DBCO）功能化的 DNA 修饰到纳米 MOFs 的表面，从而实现了核酸与 MOFs 的有效偶联。与相同尺寸（14 nm 和 19 nm）的非功能化 MOFs 纳米颗粒相比，合成的核酸-MOFs 偶联物在不使用转染剂的情况下，展现出了更高的胶体稳定性和增强的细胞摄取能力。这一优势使得该偶联物在生物医药应用中更具潜力。此外，研究还进一步证实，这些纳米颗粒偶联物能够以序列特异性的方式与互补的核酸进行杂交。这一特性为它们在细胞内基因调控等领域的应用提供了广阔的前景。这一研究成果将为 MOFs 在生物医药领域的应用开辟新的道路，为未来的研究提供新的思路和方向。

Tang 等报道中提到，ZIF-8 被用于质粒 DNA（pDNA）的传递研究[86]。为了提高其负载能力和与 pDNA 的结合能力，研究者采用聚乙烯亚胺（PEI）对 ZIF-8 进行了包裹处理。实验结果显示，在 MCF-7 细胞中，经过优化的 ZIF-8 载体展现出了良好的基因传递和表达效果。另外，周翔等研究者构建了一系列孔径逐渐增大的 IRMOF 材料，从 NIR-MOF-74-Ⅱ至 NIR-MOF-74-Ⅴ，其孔径从 2.2 nm 扩展至 4.2 nm[87]。他们利用这些材料的孔隙空间，实现了对单链 DNA（ssDNA）的可控精准负载。通过同步辐射 X 射线衍射技术的验证，证实 ssDNA 确实分布于 MOFs 晶格之中。这种包封在孔隙中的 ssDNA 在生物体内循环过程中得到了良好的保护，有效防止了降解和失活。值得一提的是，由于 MOFs 与 DNA 分子之间的主客体作用力相对较弱，弱于氢键，这使得复合物在进入细胞后能够更容易地与互补链段配对，从而实现高效释放。实验进一步显示，NIR-MOF-74-II 至 NIR-MOF-74-III 体系在小鼠和人体多种免疫细胞中均实现了优异的转染效果，其效率甚至高于商用试剂，且毒性更低。这一研究成果为 MOFs 在基因传递和表达领域的应用提供了新的可能性，并有望为未来的生物医药研究带来突破。

6.4.2 负载蛋白质

蛋白质是由长链的氨基酸组成的大分子，它们具有大量的功能如 DNA 复制、代谢反应催化作用和分子运输。蛋白质因其庞大的尺寸、带电的表面以及对环境的高度敏感性，在试图自然穿越细胞膜时，往往难以保持其结构的完整性。这使得蛋白质在不损失自身结构特性的前提下通过细胞膜成为一项极具挑战性的任务[88]。由于高生物活性和特异性，蛋白质药物在诊断和治疗方面都取得了巨大的成功[89]。将活性蛋白质直接递送至目标的想法正变得越来越有吸引力。然而，由于其体积大和环境敏感性，蛋白质不容易在细胞外液中保持生物活性，还能不受损伤地穿过细胞膜[89]。

研究者已经开发了一系列技术和载体来递送蛋白质，所以蛋白质药物有望治疗包括癌症在内的多种复杂疾病。蛋白质药物应用的首要任务是将大量生物活性蛋白质精确递送至肿瘤部位，由于载体和蛋白质之间的非共价相互作用，金属有机框架（MOFs）被广泛认为是封装蛋白质药物的有前途的载体。

Mao 等[89]开发了 ZIF-90 金属有机框架材料作为一个通用平台，即为将不同的蛋白质传递到细胞质中，而不依赖于它们的大小和分子量。通过咪唑-2-羧醛、Zn^{2+}和蛋白质的自组装来进行蛋白包封在 ATP 存在的情况下，可以观察到纳米颗粒的降解以释放蛋白质。Yang 等[90]报告了 ATP 响应性沸石咪唑框架 90（ZIF-90）已被开发为一种通用的平台，可用于细胞溶质蛋白递送和 CRISPR/Cas9 基因组编辑。这一系统利用咪唑-2-甲醛、Zn^{2+}与蛋白质之间的自组装过程，形成稳定的 ZIF-90/蛋白质纳米颗粒，从而有效地封装和保护蛋白质。研究发现，在 ATP 的存在下，ZIF-90 的结构会发生变化。ATP 与 ZIF-90 中的 Zn^{2+}存在竞争配位关系，这导致 ZIF-90/蛋白质纳米颗粒的降解，并进而释放所封装的蛋白质。这一机制为蛋白质在细胞内的按需释放提供了可能。进一步的细胞内递送研究表明，ZIF-90/蛋白质纳米颗粒具有出色的通用性。无论蛋白质的大小和分子量如何，这些纳米颗粒都能成功地将多种蛋白质递送到细胞质中，为细胞生物学研究和医学应用提供了新的工具和策略。Chen 等报道了一种通用的与表面电荷无关的 MOFs 材料负载蛋白质的方法[91]。蛋白质周围核前簇的加速形成是这一方案的成功关键，包括酶等 12 种具有不同表面化学性质的蛋白质被成功封装到 ZIF-8 中，这个过程与蛋白质的表面电荷无关。该研究通过一个简单的 pH 修饰，实现无损封装剂的可控释放。

酶，作为一类具备高度选择性的蛋白质，能够催化生物体内众多复杂反应。迄今为止，酶与金属有机框架（MOFs）的复合材料在催化、传感及检测领域均得到了广泛而深入的研究。最近，Zhou 研究小组发表了关于 MOFs 纳米载体在癌症治疗中的创新应用，即通过细胞传递酶来实施治疗。相较于传统的化疗方法，这些 MOFs 展现出了更高的选择性和更低的全身毒性，为癌症治疗提供了新的可能性[92]。在 Zhou 小组的研究中，酪氨酸酶被巧妙地封装在 PCN-333（Al）中，形成了酶-MOF 纳米反应器。这一反应器能够激活癌症前药物扑热息痛，为其在癌症治疗中的应用开辟了新路径。这一研究成果不仅丰富了酶与 MOFs 复合材料的应用领域，也为癌症治疗提供了新的策略和方法。

Jana 等用蛋白质封闭的 Zr 基 MOF（PCN-224）的纳米粒子可用于高效靶向癌症治疗[93]。未修饰的 PCN-224 表面预涂有谷胱甘肽转移酶（GST）融合的可靶向亲和体（GST-Afb）蛋白，通过简单的混合缀合而不是使诱导蛋白质损伤的化学修饰。

Sheng 等报告了一种基于 MOFs 的蛋白质递送系统[94]，该系统能够通过深层组织可穿透的近红外（NIR）光精确控制蛋白质释放。通过将目标蛋白和上转换纳米粒子（UCNP）封装在沸石咪唑酯框架（ZIF-8）MOF 中，并进一步将光酸生成剂（PAG）捕获在 ZIF-8 的孔隙中构建了响应系统。在 NIR 光照射下，UCNP 发出紫外光激活 PAG 产生质子使得局部 pH 酸化，从而使 ZIF-8 降解以实现时空控制的蛋白质释放。使用胰岛素作为模型蛋白质，证明了该系统允许按需控制其使用 NIR 光对糖尿病的治疗效果。此外，Liu 等[95]开发了一种通过蛋白质-金属离子-有机配体协调靶向肿瘤的胶原酶封装 MOFs，用于改进深部组织胰腺癌光免疫疗法。通过精确控制金属离子、蛋白质组氨酸残基和配体比例的便捷方法实现了超高包封率（质量分数 80.3%），可释放对肿瘤细胞外基质具有高酶活性的胶原酶达到肿瘤微环境后的调节。通过将热疗剂与强近红外吸收（1064 nm）相结合，可以诱导急性免疫原性，从而激活宿主免疫并产生全身免疫记忆，从而预防肿瘤的发展和复发。

Tokitaka Katayama 组最近报道了特定的 MOF 纳米颗粒可以吸附和保留蛋白质[96]，这表明 MOFs 纳米颗粒可能具有作为新型细胞培养支架的优势。然而，MOFs 纳米颗粒不能用作细胞的二维支架。因此，Katayama 等建立了一种自下而上的技术来在聚合物薄膜上构建二维 MOF[MIL-53(Al)]。开发的二维 MIL-53(Al)薄膜[fMIL-53(Al)]表现出高血清蛋白吸附、保留、与传统的细胞培养支架相比，具有补充能力。用作模型蛋白的β-半乳糖苷酶吸附在 MIL-53(A1)上表现出原始的酶活性，表明蛋白质在吸附过程中没有变性。

由于 MOFs 在水中容易团聚和塌陷，在金属有机框架（MOFs）上受控负载多种活性物质并保持单分散性长期以来一直是一个巨大的挑战。在此，Gong 等开发了一种新兴的自稳定双酶递送系统[97]，具有长时间单分散性，无需任何化疗即可实现协同癌症饥饿和光热治疗（PTT）。葡萄糖氧化酶（GOx）的负载赋予了与 MOFs 的强配位相互作用，这不仅在没有任何表面修饰的情况下赋予了强大的自稳定单分散性，而且还保证了具有催化和光热转换特性的 Au 纳米粒子的 GOx 类纳米酶的均匀沉积。他们的设计建立了一个有效的协同治疗平台来产生活性氧、降低线粒体膜电位并诱导细胞凋亡。免疫荧光研究证实激活凋亡途径的 Caspase 3 蛋白，并显著抑制癌细胞的增殖。

Abuçafy 等通过低温喷雾干燥途径从 MIL-100（Fe）的纳米颗粒聚集体中开发出多孔胶囊[98]。这使得在 MOF NPs 的孔内同时一锅封装高负载的抗肿瘤药物甲氨蝶呤，在特定介孔腔内装载胶原酶（COL），在胶囊形成时，增强肿瘤治疗。与裸露的 MOF NP 相比，这种关联在模拟体液条件下可以更好地控制活性部分、MTX 和胶原酶的释放。此外，装载的 MIL-100 胶囊对 A-375 癌细胞系的选择性毒性比正常 HaCaT 细胞高 9 倍，表明 MTX@COL@MIL-100 胶囊可能在选择性治疗癌细胞方面具有潜在应用。

Kim 等[99]报告介绍了一种创新的载药系统，该系统结合了透明质酸（HA）和 PCN-224 nanoMOF 的酶反应性，以及聚合物涂层的 MOF 技术（图 6.4）。在这个系统中，纳米 MOF 的外表面通过 Zr 团簇与 HA 的羧酸之间的多价配位键，被 HA 稳定地覆盖，形成了一个有效的“开关”机制。这种设计使得药物载体能够在 CD44 过表达的癌细胞中选择性地积聚。一旦进入癌细胞环境，HA 涂层会响应并释放酶反应性药物，从而实现精准治疗。作为药物载体的 PCN-224 nanoMOF，凭借其固有的特性，能够促进药物更稳定地向癌细胞转移。更重要的是，这种系统还允许结合光动力学治疗，为癌症治疗提供了更多可能性。综上所述，这种结合了透明质酸、PCN-224 nanoMOF 以及聚合物涂层的载药系统，不仅提高了药物递送的准确性和稳定性，还为癌症的综合治疗提供了新的思路和方法。

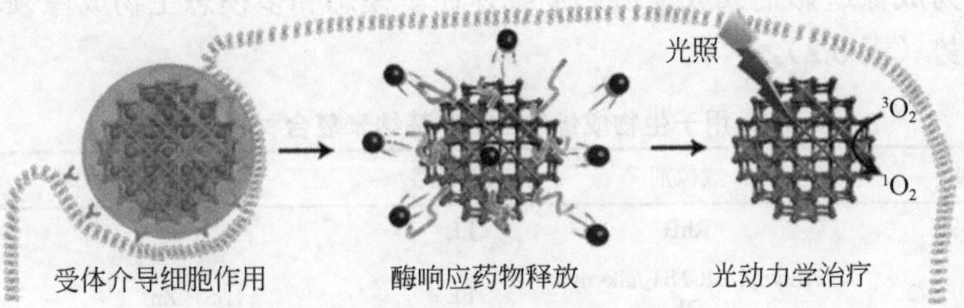

图 6.4　酶反应 MOF 载药系统[99]

覆有金属有机框架（MOFs）的 Janus 载体细胞是通过用细胞毒性酶不对称地固定基于锌的 MOFs 纳米颗粒而产生的，该细胞毒性酶被内部封装在载体细胞的表面。通过维持常规活细胞的生物学和结构特征，其研究中[100]开发的涂有 MOFs 的 Janus 细胞保留了细胞与其微环境的固有结合能力，加载到 Janus 细胞另一面上的互连 MOFs 无法穿透该细胞。因此，保护了载体细胞免受 MOFs 中包含的细胞毒性药物的侵害。当在酸性环境中从 MOFs 纳米颗粒中释放蛋白酶 K 的化学治疗蛋白时，这些 MOF-Janus 载体细胞可成功消除三维（3D）肿瘤球体。

6.5　生 物 成 像

生物成像是指在细胞、组织或生物体水平上直观地捕捉生物过程的技术。它包括一系列方法，如显微镜[101]、荧光成像[102]、MRI[103]和 CT 扫描[104]等以提供对生物系统结构、功能和动力学的详细见解。生物医学成像技术的发展和成熟为各种疾病的诊断提供了极大的便利。满足成像和药物传递的多功能药物载体已成为癌症治疗的一个重要的新方向。成像剂可以在靶组织中产生信号或增强信号对比度，可作为一种诊断工具。该领域对于医学诊断、了解疾病机制和评估治疗效果、促进生物学和医学的研究和开发是至关重要的。使用 MOFs 进行生物成像可显著增强在细胞或组织水平上可视化生物过程的能力，从而实现靶向药物递送和治疗监测。此应用程序利用 MOFs 的结构多样性和可调性来整合显像剂，提供高分辨率图像。它提供了一种非侵入性方法来跟踪治疗药物的生物分布，实时评估治疗效果，并促进个性化医疗策略的制定[105]。

MOFs 基纳米复合材料因其独特的性质和多样的应用功能，在多个医学成像技术中得到了广泛应用。这包括荧光成像（FL）[106]、计算机断层扫描（CT）[107]、磁共振成像（MRI）[108]以及正电子发射断层扫描（PET）[109]等多种成像方式。这些成像技术在医学诊断、疾病监测以及治疗效果评估等方面发挥着重要作用，而 MOFs 基纳米复合材料的引入，为这些技术提供了更为高效、精确的成像手段，有助于推动医学诊疗技术的进一步发展。在本节中，将介绍基于 MOFs 的纳米复合材料作为成像造影剂或成像造影剂载体在单模态和多模态生物成像领域的应用和潜在优势（表 6.2）。

表 6.2　用于生物成像的 MOF 基纳米复合材料的例子

MOFs	成像剂	成像策略	测试细胞	参考文献
ZIF-90	RhB	FL	ATP	[110]
TP-MOF 1/2	Alkynyl-BR-NH_2/alkynyl-DL	FL	H_2S，Zn^{2+}	[111]
Bi-NU-901	Bi-NU-901	CT	–	[112]
MIL-USPIO-cit	γ-Fe_2O_3@MIL-100	MRI	小鼠腹部细胞	[113]
Fe_3O_4@UiO-66@WP6	Fe_3O_4	MRI	HeLa 细胞	[114]
89Zr-UiO-66/Py-PGA-PEG-F3	Zr-UiO-66	PET	MDA-MB-231 tumor-bearing mice	[115]
Au@Ag NRs4-ATP@ZIF-8	4-ATP	SERS	f HeLa，MCF-7，LNCaP，QGY-7703，HCT116 and MDA-MB-231 cells	[116]
UMP-FA	UCNPs&Fe-MIL-101-NH_2	FL&MRI	KB tumor-bearing mice	[117]
Fe^{2+}-adsorbed ZIF-8	ZnO，Fe_3O_4	FL&MRI	U87 xenograft tumor mice	[118]
Au@MIL-88（Fe）	Au nanorods，MIL-88（Fe）	CT&MRI&PAI	U87 MG-orthotopic tumor-bearing mice	[119]
MOF@HA@ICG	ICG，MIL-100（Fe）	FL&PAI&MRI	MCF-7 cells/xenograft tumors	[120]
Gd/Yb-MOFs-Glu	DOX，Gd/Yb-MOFs	FL&CT&MRI	HeLa tumor-bearing mice	[121]

镧系 MOFs 如 Tb-MOFs、Yb-MOFs 和 Eu-MOFs 具有良好的光稳定性和特征荧光发射，由于π共轭有机连接体在与镧系金属紧密连接时，作为“天线”极大地提高了镧系的特征荧光发射和稳定性。近年来，镧系金属基 MOFs 利用其可调的孔隙率和稳定的荧光特性，成为一种生物成像和纳米医学应用的新型生物材料。

6.5.1　荧光成像

荧光成像（fluorescence imaging，FL）是一种基于荧光物质对特定波长光的吸收以及随后发射不同波长光的性质来进行的成像技术。当荧光分子吸收特定波长的激发光后，会以较长的波长发射光然后被探测并用于成像。荧光成像在生物医学研究和临床诊断中被广泛应用，特别是在细胞标记、生物分子追踪、疾病诊断和治疗监测等方面。

用荧光分子或量子点功能化的 MOFs 可用于跟踪药物递送途径，并实时监测生物系统内治疗剂的释放。MOFs 的高表面积允许掺入大量的荧光标签，从而增强了它们在生物成像工具下的可见性。FL 作为一种普遍使用的成像方法，通过光学分子获取生物分布和含量的可视化信息，具有灵敏度高、分辨率好、操作方便等优点。如用于跟踪生物系统中药物递送的荧光基团功能化 MOFs 是用荧光染料（如罗丹明或荧光素）功能化。这些荧光标签允许在显微镜下或通过成像技术观察 MOFs，从而可以实时监测细胞或组织内封装药物的递送和释放。这种方法将 MOFs 的高负载能力和可调性与荧光标记物提供的可见性相结合，以实现有效的生物成像应用。

Mao 和同事构建了一个简单的由 Zn^{2+}和咪唑-2-羧基醛配体组成基于 MOFs 的荧光探针 RhB/ZIF-90，用于活细胞中线粒体三磷酸腺苷（ATP）的荧光成像检测[110]。将罗丹明（RhB）封装在 ZIF-90 中，由于其自猝灭效应，其荧光被猝灭。在 ATP 存在的情况下，RhB/ZIF-90 被分解，导致 RhB 的释放和荧光恢复，用于细胞内 ATP 成像。

Zhang 和同事首次报道了基于双光子 MOF 的荧光探针分别对活细胞和组织中的硫化氢和 Zn^{2+}进行生物成像（图 6.5）[122]。用两种有机探针（硫化氢的炔基-br-NH_2 和 Zn^{2+}）对 PCN-58 表面进行修饰，分别形成 TP-MOF 探针 1 和 TP-MOF 探针 2。实验结果表明，TP-MOF 探针在大鼠肝组织中表现出优异的光稳定性、优异的选择性、良好的生物相容性以及理想的穿透深度（最高可达 130 μm），这为使用 MOFs 检测生物组织中的活性成分提供了一种方法。

Kyeng 小组[123]把生物相容性聚合物表面活性剂（F127）和 Tb-MOF 进行机械研磨，制备了具有良好胶体稳定性和稳定荧光性能的 Tb 基金属有机框架纳米颗粒（Tb-MOF NPs）。镧系元素的特征荧光特性使这种纳米材料可以作为细胞成像探针。利用 Tb-MOF NPs 的多孔性成功地装载 DOX 并递送，杀死癌细胞。水中的 Tb-MOF 在 488、541、584 和 620 nm 处均有荧光发射，且稳定了一个多月。

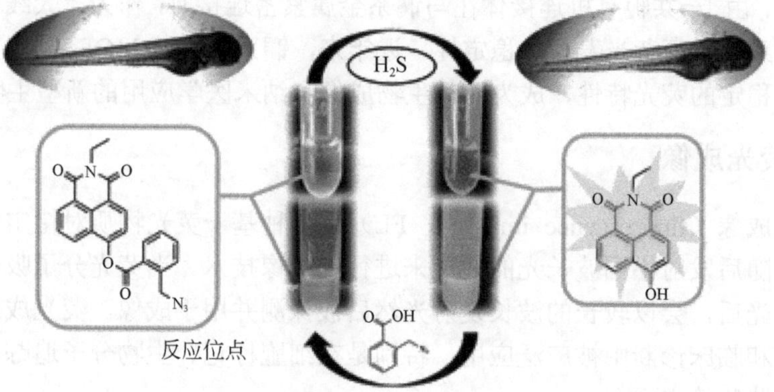

图 6.5　MOF 的荧光探针对活细胞和组织中的硫化氢和 Zn^{2+}进行生物成像[122]

Zhang 小组以 MOF-74（Zn）为模板，将乙二胺嫁接在开放金属位点上，通过 MOF 衍生物荧光增强检测 TBBPA，开发了一种高选择性和灵敏度的荧光检测器，用于快速测定四溴双酚 A（TBBPA）[124]。通过表征显示了理想的光致发光性能。在 50～400 μg/L 范围内，荧光增强效果与 TBBPA 的浓度呈良好的线性关系，其检测限可达 0.75 μg/L。此外，荧光增强的可能的传感机制也属于福斯特共振能量转移（FRET）。研究结果为 TBBPA 的检测提供了一种方便而快速的方法。

Zhao 等提出了通过发光猝灭机制起作用的一种使用镧系元素 MOF $Eu_2(sbdc)_3(H_2O)_3$(Eu-sbdc)检测 ATP 的新方法[125]。该方法具有较高灵敏度和快速检测能力，最重要的是在环境监测和医疗保健应用中利用镧系元素 MOFs 的独特特性创造一种用于 ATP 检测的灵敏和选择性发光传感器，对诊断与 ATP 水平异常相关的疾病或环境危害具有潜在意义。

6.5.2　磁共振成像

磁共振成像（magnetic resonance imaging，MRI）是一种具有良好空间分辨率的无创诊断技术，已成为详细可视化人体内部结构的最重要的临床工具之一，许多研究者致力于开发新型的药物载体希望作为 MRI 造影剂进行疾病的早期检测。

含有氧化铁等磁性纳米颗粒的 MOFs 可用作 MRI 中的造影剂的材料增强了图像的对比度，提供了有关治疗剂在体内分布的详细信息，并能够对药物递送过程进行非侵入性监测。

Steunou 及其团队精心研发了一种具备超顺磁性的介孔 MIL-100（Fe）复合材料，其特点在于表面装饰有磁赤铁矿（γ-Fe_2O_3）纳米颗粒（NPs），这一创新设计使得该复合材料在 MRI 成像与癌症治疗领域展现出巨大的潜力[126]。在生理条件下，这种复合材料展现出卓越的稳定性和 MRI 性能。特别值得一提的是，当γ-Fe_2O_3 NPs 的质量分数达到 10%时，其相对弛豫值竟然是 MIL-100（Fe）的 9 倍，这一显著提升可归因于复合材料所具备的高饱和磁化强度。此外，经过 DOX 负载后的这种复合材料不仅展现出良好的生物相容性，还表现出较高的抗肿瘤活性。因此，它有望成为体内 MRI 成像的理想造影剂，并在药物递送系统（DDS）中发挥关键作用，为癌症治疗带来新的希望。

Yang 小组制备的 Fe_3O_4@MOF 核壳杂化体，采用 UiO-66-NH_2 壳体在四氧化三铁核表面的原位生长方法，具有优越的 MRI 和磁分离能力。通过与修饰基团的主客体相互作用，将 WP6 引入 5-Fu 负载的 Fe_3O_4@MOF 杂化体，其具有紧密性可调节纳米阀表面[127]。

MOFs 的磁共振成像也可用于阿尔茨海默病的辅助诊断。选择磁性纳米材料 Fe-MIL-88-NH_2 包封亚甲基蓝（MB，一种 tau 聚集抑制剂），并将其作为磁共振造影剂。随后，将靶向试剂 5-amino-3-(pyrrolo[2,3-*c*]pyridin-1-yl)isoquinoline（脱氟 MK6240，DMK6240）通过 1,4,7-三氮环甲烷-1,4,7-三乙酸（NOTA）连接到 Fe-MIL-88B-NH_2 表面，增强过磷酸化 tau 靶向性形成先进的给药系统（Fe-MIL-88B-NH_2-NOTA-DMK6240/MB）。Fe-MIL-88B-NH_2-NOTA-DMK6240/MB 的表面特性使其具有出色的磁共振成像能力，并在体内和体外改善 AD 症状[128]。

林彩雪等开发了复合材料 MOFs/CDs@OCMC[129]，在 *O*-羧甲基壳聚糖（OCMC）壳内集成封装碳点（CD）和 MIL-100 金属有机框架（MOFs），旨在提供具有双重成像模式的生物相容性、高效的药物递送系统（图 6.6）。他们所开发

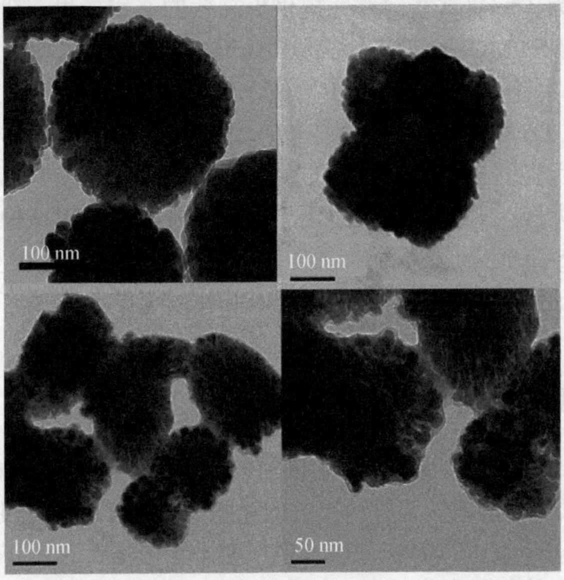

图 6.6　壳内集成封装碳点（CD）和 MIL-100 的 TEM[129]

的 MOFs/CDs@OCMC 纳米颗粒能够将 FOI 和 MRI 成像与受控的 pH 响应性药物释放相结合，为早期癌症检测与靶向治疗相结合提供了一种新的方法，并将有可能提高癌症治疗的疗效和安全性。

6.5.3 计算机断层扫描成像

计算机断层扫描（computed tomography，CT）成像是一种医学影像技术，通过使用 X 射线和计算机处理生成人体内部的横截面图像。与传统的 X 射线不同，CT 能够提供更详细的结构信息，包括组织、器官和骨骼的精细图像，有助于医生进行诊断、监测和治疗计划的制定。CT 成像的关键在于对 X 射线的吸收和散射能力，这通常需要使用对比剂来增强。对比剂是一种可以被体内特定部位吸收或集中的物质，以提高成像质量。然而，传统对比剂如碘化合物不仅可能会引起某些患者的不良反应，其在体内的分布和清除速度也限制了成像的效果。

MOFs 中可具有高 X 射线衰减系数的金属离子，如铋或金，可用作 CT 成像中的造影剂。Lee Robison 等研究了一种基于铋 Bi-NU-901 新的 CT 造影剂 MOFs。与传统的碘和钡剂相比，由于铋的原子序数更高且无毒，可显著改善 X 射线衰减，该应用程序能够可视化 MOFs 在药物递送应用中的生物分布和定位，从而深入了解其性能和功效。CT 是一种通过扫描组织间的差异构建的三维灰度图像，为医生提供了清晰展现组织内部结构的视觉依据，从而有力地支持了疾病的诊断工作。然而，目前临床上广泛使用的碘、钡等小分子造影剂，它们在体内的分布不理想、清除速度快以及需要极高剂量（甚至达到数十克）等缺陷，使得它们的应用受到了很大的限制。为了解决这些问题并提供更高效的造影效果，研究者们开始关注基于金属有机框架（MOFs）的造影剂。Farha 及其团队成功合成了一种新型的 MOF 造影剂，它由铋（Bi）-NU-901 构成，其结构包括$[Bi_6O_4(OH)_4(NO_3)_6(H_2O)]$（$H_2O$）节点和 H_4TBAPy 配体。这种 MOFs 造影剂不仅克服了传统小分子造影剂的种种不足，而且能够提供高对比度的图像增强效果，为 CT 诊断的精确性和可靠性带来了新的突破[112]。Li 等开发了用于表面增强拉曼散射成像和药物递送的生物相容性 Au@Ag @ZIF-8 核壳纳米颗粒[116]，可通过表面增强拉曼散射（SERS）进行成像进而用于癌细胞成像和药物递送，并使用 ZIF-8 金属有机框架材料进行载药。

6.5.4 正电子发射断层扫描

相较于其他成像技术，正电子发射断层扫描（positron emission tomography，PET）成像凭借其快速的成像速度、出色的灵敏度、深厚的穿透度以及强大的定量能力而备受瞩目。Hong 和同事在其研究中报道了一种基于 NMOF 的纳米平台[115]，该平台本质上具有放射性标记特性，特别适用于三阴性乳腺肿瘤的 PET 成像。该平台以 Zr-UiO-66 作为金属聚簇，通过 1,4-苯二羧酸和苯甲酸的连接形成稳定结构。在装载抗癌药物 Dox 后，平台进一步经过聚乙二醇（Py-PGA）和 PEG 的功能化，以增强 NMOF 的稳定性和活性靶向性。

值得一提的是，该平台不仅具备高负荷 Dox 的能力，使 Dox 作为抗肿瘤治疗的有效药物，而且 Dox 还能作为荧光可视化器[130]，为研究者提供额外的诊断信息。此外，Zr 的长半衰期优势（$t_{1/2}$=78.4 h）使得研究人员能够监测 MOFs 载药体在体内腹腔注射后长达 120 h 的分布和清除过程，为药物疗效评估和安全性评估提供了重要依据。实验结果表明，这种 MOFs 基材料展现出了作为安全稳定的 PET 成像和肿瘤治疗纳米平台的巨大潜力，为未来的医学诊断和治疗提供了新的可能性。

MOFs 通过结合不同成像模式，可以提供更全面、更准确的生物医学信息。MOFs 的高孔隙性和可调性使其能够同时载药和成像，从而实现治疗与诊断一体化。这种一体化系统在生物成像方面有很大的优势，如能够在药物递送的同时进行实时成像，并监控药物的分布、释放和疗效等。

参考文献

[1] Magiorakos A P, Srinivasan A, Carey R B, et al. Multidrug-resistant, extensively drug-resistant and pandrug-resistant bacteria: An international expert proposal for interim standard definitions for acquired resistance. Clinical Microbiology and Infection, 2012, 18: 268-281.

[2] Rhodes A, Evans L E, Alhazzani W, et al. Surviving sepsis campaign: International guidelines for management of sepsis and septic shock: 2016. Critical Care Medicine, 2017, 45: 486-552.

[3] Ayers P W. The physical basis of the hard/Soft acid/base principle. Faraday Discussions, 2007, 135: 161-190.

[4] 马彦平, 石磊, 何源. 微量元素铁、锰、硼、锌、铜、钼营养与人体健康. 肥料与健康, 2020, 47: 12-17.

[5] 王天歌, 刘轩彤, 段瑞, 等. 化妆品中常见维生素的功效及检测方法概述. 品牌与标准化, 2024, 1: 46-50.

[6] Hoppe A, Gueldal N S, Boccaccini A R. A review of the biological response to ionic dissolution products from bioactive glasses and glass-ceramics. Biomaterials, 2011, 32: 2757-2774.

[7] 史晓群, 杜希友. 铜死亡机制及相关抗癌药物. 生命的化学, 2024, 44: 225-232.

[8] 罗文博, 王昕阳, 郭衍科, 等. 微纳米铜粉的制备研究进展. 金属功能材料, 2024, 2: 12-21.

[9] Kim J S, Kuk E, Yu K N, et al. Antimicrobial effects of silver nanoparticles. Nanomedicine, 2007, 3: 95-101.

[10] Zhang Liuqin, Li Huali, Zhang Xiaohu, et al. Uv-curable pbma coating containing CuZn-MOF-74 for fouling-resistance. Microporous and Mesoporous Materials, 2024, 368: 113020.

[11] Kim D, Park K W, Park J T, et al. Photoactive MOF-derived bimetallic silver and cobalt nanocomposite with enhanced antibacterial activity. ACS Applied Materials & Interfaces, 2023, 15: 22903-22914.

[12] Lelouche S N, Albentosa-González L, Clemente-Casares P, et al. Antibacterial Cu or Zn-MOFs based on the 1,3,5-Tris-(styryl)benzene tricarboxylate. Nanomaterials, 2023, 13: 2294.

[13] Yang Ye, Wu Xizheng, He Chao, et al. Metal-organic framework/Ag-based hybrid nanoagents for rapid and synergistic bacterial eradication. ACS Applied Materials & Interfaces, 2020, 12: 13698-13708.

[14] 齐野，任双颂，车颖，等. 金属有机框架抗菌材料的研究进展. 化学学报，2020, 78: 613-624.

[15] Chiericatti C, Basilico J C, Basilico M L Z, et al. Novel application of hkust-1 metal-organic framework as antifungal: Biological tests and physicochemical characterizations. Microporous and Mesoporous Materials, 2012, 162: 60-63.

[16] Rauf A, Ye Junwei, Zhang Siqi, et al. Copper(II)-based coordination polymer nanofibers as a highly effective antibacterial material with a synergistic mechanism. Dalton Transactions, 2019, 48: 17810-17817.

[17] Li Xiangqian, Zhao Xinshuo, Chu Dandan, et al. Silver nanoparticle-decorated 2d Co-tcpp MOF nanosheets for synergistic photodynamic and silver ion antibacterial. Surfaces and Interfaces, 2022, 33: 102247.

[18] Lin Sha, Liu Xiangmei, Tan Lei, et al. Porous iron-carboxylate metal-organic framework: A novel bioplatform with sustained antibacterial efficacy and nontoxicity. ACS Applied Materials & Interfaces, 2017, 9: 19248-19257.

[19] Gallis D F S, Butler K S, Agola J O, et al. Antibacterial countermeasures via metal-organic framework-supported sustained therapeutic release. ACS Applied Materials & Interfaces, 2019, 11: 7782-7791.

[20] Huang Guohuan, Li Yanming, Qin Zhimei, et al. Hybridization of carboxymethyl chitosan with MOFs to construct recyclable, long-acting and intelligent antibacterial agent carrier. Carbohydrate Polymers, 2020, 233: 115848.

[21] Cai Ying, Guan Jingwei, Wang Wen, et al. pH and light-responsive polycaprolactone/curcumin@ZIF-8 composite films with enhanced antibacterial activity. Journal of Food Science, 2021, 86: 3550-3562.

[22] Yadav P, Kumari S, Yadav A, et al. Biocompatible drug delivery system based on a MOF

platform for a sustained and controlled release of the poorly soluble drug norfloxacin. ACS Omega, 2023, 8: 28367-28375.

[23] Bhat Z U H, Hanif S, Rafi Z, et al. New mixed-ligand Zn(II)-based MOF as a nanocarrier platform for improved antibacterial activity of clinically approved drug levofloxacin. New Journal of Chemistry, 2023, 47: 7416-7424.

[24] Luo Ting, Shakya S, Mittal P, et al. Co-delivery of superfine nano-silver and solubilized sulfadiazine for enhanced antibacterial functions. International Journal of Pharmaceutics, 2020, 584: 119407.

[25] Asadollahi T, Kazemi N M, Halajian S. Alginate-zein composite modified with metal organic framework for sulfasalazine delivery. Chemical Papers, 2024, 78: 565-675.

[26] Huang Rong, Cai Guangqing, Li Jian, et al. Platelet membrane-mamouflaged silver metal-organic framework drug system against infections caused by methicillin-esistant rstaphylococcus aureus. Journal of Nanobiotechnology, 2021, 19: 229.

[27] Wei Yucai, Chen Chaoxi, Zhai Shuo, et al. Enrofloxacin/florfenicol loaded cyclodextrin metal-organic-framework for drugdelivery and controlled release. Drug Delivery, 2021, 28: 372-379.

[28] He Zhiqiang, Yang Huan, Gu Yufan, et al. Green synthesis of MOF-mediated pH-sensitive nanomaterial agnps@ZIF-8 and its application in improving the antibacterial performance of agnps. International Journal of Nanomedicine, 2023, 18: 4857-4870.

[29] Shakya S, He Yaping, Ren Xiaohong, et al. Ultrafine silver nanoparticles: Ultrafine silver nanoparticles embedded in cyclodextrin metal-organic frameworks with grgds functionalization to promote antibacterial and wound healing application. Small, 2019, 15: 1970145.

[30] Zhang Wenshang, Wang Bingjie, Xiang Guangli, et al. Photodynamic alginate Zn-MOF thermosensitive hydrogel for accelerated healing of infected wounds. ACS Applied Materials & Interfaces, 2023, 15: 22830-22842.

[31] Tang Jingyue, Tang Gang, Niu Junfan, et al. Preparation of a porphyrin metal-organic framework with desirable photodynamic antimicrobial activity for sustainable plant disease management. Journal of Agricultural and Food Chemistry, 2021, 69: 2382-2391.

[32] Guo Wei, Wang Yanmin, Zhang Kai, et al. Near-infrared light-propelled MOF@Au nanomotors for enhanced penetration and sonodynamic therapy of bacterial biofilms. Chemistry of Materials, 2023, 35: 6853-6864.

[33] Gao Jie, Hao Lingwan, Jiang Rujian, et al. Surprisingly fast assembly of the MOF film for synergetic antibacterial phototherapeutics. Green Chemistry, 2022, 24: 5930-5940.

[34] Qian Shengxu, Song Lingjie, Sun Liwei, et al. Metal-organic framework/poly (E-Caprolactone) hybrid electrospun nanofibrous membranes with effective photodynamic antibacterial activities.

Journal of Photochemistry and Photobiology A: Chemistry, 2020, 400: 112626.
[35] Hao Lingwan, Jiang Rujian, Fan Yong, et al. Formation and antibacterial performance of metal-organic framework films via dopamine-mediated fast assembly under visible light. ACS Sustainable Chemistry & Engineering, 2020, 8: 15834-15842.
[36] Xiao Ya, Xu Mengran, Lv Na, et al. Dual stimuli-responsive metal-organic framework-based nanosystem for synergistic photothermal/pharmacological antibacterial therapy. Acta Biomaterialia, 2021, 122: 291-305.
[37] Soomro N A, Wu Qiao, Amur S A, et al. Natural drug physcion encapsulated zeolitic imidazolate framework, and their application as antimicrobial agent. Colloids and Surfaces B: Biointerfaces, 2019, 182: 110364.
[38] Ghaffar I, Imran M, Perveen S, et al. Synthesis of chitosan coated metal organic frameworks（MOFs） for increasing vancomycin bactericidal potentials against resistant s. aureus strain. Materials Science and Engineering C-Materials for Biological Applications, 2019, 105: 110111.
[39] Tang Jingyue, Ding Guanglong, Niu Junfan, et al. Preparation and characterization of tebuconazole metal-organic framework-based microcapsules with dual-microbicidal activity. Chemical Engineering Journal, 2019, 359: 225-232.
[40] Esfahanian M, Ghasemzadeh M A, Razavian S M H. Synthesis, identification and application of the novel metal-organic framework Fe_3O_4@Paa@ZIF-8 for the drug delivery of ciprofloxacin and investigation of antibacterial activity. Artificial Cells Nanomedicine and Biotechnology, 2019, 47: 2024-2030.
[41] Andre V, Silva A, Fernandes A, et al. Mg- and Mn-MOFs boost the antibiotic activity of nalidixic acid. ACS Applied Bio Materials, 2019, 2: 2347-2354.
[42] Gwon K, Han I, Lee S, et al. Novel metal-organic framework-based photocrosslinked hydrogel system for efficient antibacterial applications. ACS Applied Materials& Interfaces, 2020, 12: 20234-20242.
[43] Yao Shun, Chi Junjie, Wang Yuetong, et al. Zn-MOF encapsulated antibacterial and degradable microneedles array for promoting wound healing. Advanced Healthcare Materials, 2021, 10: 2100056.
[44] Rezaee R, Montazer M, Mianehro A, et al. Single-step synthesis and characterization of Zr-MOF onto wool fabric: Preparation of antibacterial wound dressing with high absorption capacity. Fibers and Polymers, 2022, 23: 404-412.
[45] Sanaei-Rad S, Ghasemzadeh M A, Aghaei S S. Synthesis and structure elucidation of $ZnFe_2O_4$/IrMOF-3/Go for the drug delivery of tetracycline and evaluation of their antibacterial activities. Journal of Organometallic Chemistry, 2022, 960:122221.
[46] Slouma M, Lahmar W, Mohamed G, et al. Associated factors with liver fibrosis in rheumatoid

arthritis patients treated with methotrexate. Clinical Rheumatology, 2024, 43: 929-938.

[47] Guo Lina, Chen Yang, Wang Ting, et al. Rational design of metal-organic frameworks to deliver methotrexate for targeted rheumatoid arthritis therapy. Journal of Controlled Release, 2021, 330: 119-131.

[48] Li Zhen, Zhao Songjian, Wang Huizhen, et al. Functional groups influence and mechanism research of UiO-66-Type metal-organic frameworks for ketoprofen delivery. Colloids and Surfaces B: Biointerfaces, 2019, 178: 1-7.

[49] Li Zhen, Li Zhenjian, Li Sijing, et al. Potential application development of Sr/Hcooh metal organic framework in osteoarthritis. Microporous and Mesoporous Materials, 2020, 294: 109835.

[50] Guo Lina, Zhong Shenghui, Liu Peng, et al. Radicals scavenging MOFs enabling targeting delivery of sirna for rheumatoid arthritis therapy. Small, 2022, 18: 2202604.

[51] Xue Song, Zhou Xiaojun, Sang Weilin, et al. Cartilage-targeting peptide-modified dual-drug delivery nanoplatform with nir laser response for osteoarthritis therapy. Bioactive Materials, 2021, 6: 2372-2389.

[52] Taherzade S D, Rojas S, Soleimannejad J, et al. Combined cutaneous therapy using biocompatible metal-organic frameworks. Nanomaterials, 2020, 10: 2296.

[53] Zhao Heng, Saad S, Mielcarek A M, et al. Hierarchical superparamagnetic metal-organic framework nanovectors as anti-inflammatory nanomedicines. Journal of Materials Chemistry B, 2023, 11: 3195-3211.

[54] Villarreal-Ramos B, Manser J M, Collins R A, et al. Cattle immune responses to tetanus toxoid elicited by recombinant s. typhimurium vaccines or tetanus toxoid in alum or freund's adjuvant. Vaccine, 2000, 18: 1515-1521.

[55] Li Congcong, Chen Chaoxi, Wei Yucai, et al. Cyclodextrin metal-organic framework as vaccine adjuvants enhances immune responses. Drug Delivery, 2021, 28: 2594-2602.

[56] Zheng Jiewei, Solomon M B, Rawal A, et al. Passivation-free, liquid-metal-based electrosynthesis of aluminum metal-organic frameworks mediated by light metal activation. ACS Nano, 2023, 17: 25532-25541.

[57] Guo Hongxu, Zhang Yanhui, Zheng Zishan, et al. Facile one-pot fabrication of Ag@MOF(Ag) nanocomposites for highly selective detection of 2,4,6-trinitrophenol in aqueous phase. Talanta, 2017, 170: 146-151.

[58] Miao Yangbao, Pan Wenyu, Chen Kuanhung, et al. Engineering a nanoscale Al-MOF-armored antigen carried by a “trojan horse” -like platform for oral vaccination to induce potent and long-lasting immunity. Advanced Functional Materials, 2019, 29: 1904828.

[59] Kang S W, Choi S H, Jeong J H, et al. Difference of quantitative EEG between Alzheimer’s

disease (AD) dementia and non-dementia Ad. Alzheimer's & Dementia, 2020, 16: e044300.
[60] Gharat R, Dixit G, Khambete M, et al. Targets, trials and tribulations in Alzheimer therapeutics.European Journal of Pharmacology, 2024, 962: 176230.
[61] Yu Dongqin, Guan Yijia, Bai Fuquan, et al. Metal-organic frameworks harness Cu chelating and photooxidation against amyloid b aggregation in vivo. Chemistry - A European Journal, 2019, 25: 3489-3495.
[62] Zhao Jinhua, Yin Fucheng, Ji Limei, et al. Development of a tau-targeted drug delivery system using a multifunctional nanoscale metal-organic framework for Alzheimer's disease therapy. ACS Applied Materials & Interfaces, 2020, 12: 44447-44458.
[63] Santos J, Quimque M T, Liman R A, et al. Computational and experimental assessments of magnolol as a neuroprotective agent and utilization of UiO-66(Zr) as its drug delivery system. ACS Omega, 2021, 6: 24382-24396.
[64] Yang Xiaoxi, Feng Pengfei, Cao Jing, et al. Composition-engineered metal-organic framework-based microneedles for glucose-mediated transdermal insulin delivery. ACS Applied Materials & Interfaces, 2020, 12: 13613-13621.
[65] Baldassarro V A, Lorenzini L, Giuliani A, et al. Molecular mechanisms of skin wound healing in non-diabetic and diabetic mice in excision and pressure experimental wounds. Cell and Tissue Research, 2022, 388: 595-613.
[66] Yin Mengting, Wu Jiayingzi, Deng Mingwu, et al. Multifunctional magnesium organic framework-based microneedle patch for accelerating diabetic wound healing. ACS Nano, 2021, 15: 17842-17853.
[67] Sun Yi, Bao Bingbo, Zhu Yu, et al. An fps-zm1-encapsulated zeolitic imidazolate framework as a dual proangiogenic drug delivery system for diabetic wound healing. Nano Research, 2022, 15: 5216-5229.
[68] Labiris N R, Dolovich M B. Pulmonary drug delivery art I: Hysiological factors affecting therapeutic effectiveness of aerosolized medications. British Journal of Clinical Pharmacology, 2003, 56: 588-599.
[69] Togami K, Kanehira Y, Yumita Y, et al. Heterogenous intrapulmonary distribution of aerosolized model compounds in mice with bleomycin-induced pulmonary fibrosis. Journal of Aerosol Medicine and Pulmonary Drug Delivery, 2023, 36: 289-299.
[70] Lechanteur A, Evrard B. Influence of composition and spray-drying process parameters on carrier-free dpi properties and behaviors in the lung: A review. In Pharmaceutics, 2020, 12: 55.
[71] Gardikis K, Hatziantoniou S, Bucos M, et al. New drug delivery nanosystem combining liposomal and dendrimeric technology (liposomal locked-in dendrimers) for cancer therapy. Journal of Pharmaceutical Sciences, 2010, 99: 3561-3571.

[72] Fernández-Paz E, Feijoo-Siota L, Gaspar M M, et al. Microencapsulated chitosan-based nanocapsules: A new platform for pulmonary gene delivery. In Pharmaceutics, 2021, 13: 1377.

[73] Kuehl P J, Chand R, McDonald J D, et al. Pulmonary and regional deposition of nebulized and dry powder aerosols in ferrets. AAPS PharmSciTech, 2019, 20: 242.

[74] Jarai B M, Stillman Z, Attia L, et al. Evaluating UiO-66 metal-organic framework nanoparticles as acid-sensitive carriers for pulmonary drug delivery applications. ACS Applied Materials & Interfaces, 2020, 12: 38989-39004.

[75] Li Haiyan, Zhu Jie, Wang Caifen, et al. Paeonol loaded cyclodextrin metal-organic framework particles for treatment of acute lung injury via inhalation. International Journal of Pharmaceutics, 2020, 587: 119649.

[76] Zhou Yixian, Zhao Yiting, Niu Boyi, et al. Cyclodextrin-based metal-organic frameworks for pulmonary delivery of curcumin with improved solubility and fine aerodynamic performance. International Journal of Pharmaceutics, 2020, 588: 119777.

[77] Tan Xuyu, Jia Fei, Wang Ping, et al. Nucleic acid-based drug delivery strategies. Journal of Controlled Release, 2020, 323: 240-252.

[78] Cox A J, Bengtson H N, Rohde K H, et al. DNA nanotechnology for nucleic acid analysis: Multifunctional molecular DNA machine for Rna detection. Chemical Communications, 2016, 52: 14318-14321.

[79] Cutler J I, Zhang Ke, Zheng Dan, et al. Polyvalent nucleic acid nanostructures. Journal of the American Chemical Society, 2011, 133: 9254-9257.

[80] Shrivastava G, Bakshi H A, Aljabali A A, et al. Nucleic acid aptamers as a potential nucleus targeted drug delivery system. Current Drug Delivery, 2020, 17: 101-111.

[81] Ma Yanyan, Yin Junling, Li Guanghan, et al. Simultaneous sensing of nucleic acid and associated cellular components with organic fluorescent chemsensors. Coordination Chemistry Reviews, 2020, 406: 213144.

[82] Tao Tongxiang, Rehman S, Xu Shuai, et al. A biomimetic camouflaged metal organic framework for enhanced sirna delivery in the tumor environment. Journal of Materials Chemistry B, 2024, 12: 4080-4096.

[83] LeRoy M A, Perera A S, Lamichhane S, et al. Colloidal stability and solubility of metal-organic framework particles. Chemistry of Materials, 2024, 36: 3673-3682.

[84] Wang Shunzhi, McGuirk C M, Ross M B, et al. General and direct method for preparing oligonucleotide-functionalized metal-organic framework nanoparticles. Journal of the American Chemical Society, 2017, 139: 9827-9830.

[85] Morris W, Briley W E, Auyeung E, et al. Nucleic acid-metal organic framework（MOF）nanoparticle conjugates. Journal of the American Chemical Society, 2014, 136: 7261-7264.

[86] Li Yantao, Zhang Kai, Liu Porun, et al. Encapsulation of plasmid DNA by nanoscale metal-organic frameworks for efficient gene transportation and expression. Advanced Materials, 2019, 31: 1901570.

[87] Peng Shuang, Bie Binglin, Sun Yangzesheng, et al. Metal-organic orameworks for precise inclusion of single-stranded DNA and transfection in immune cells. Nature Communications, 2018, 9: 1293.

[88] Wang Shunzhi, Chen Yijing, Wang Shuya, et al. DNA-Functionalized metal-organic framework nanoparticles for intracellular delivery of proteins. Journal of the American Chemical Society, 2019, 141: 2215-2219.

[89] Guo Yan, Li Yantao, Zhou Sijie, et al. Metal-organic framework-based composites for protein delivery and therapeutics. ACS Biomaterials Science & Engineering, 2022, 8: 4028-4038.

[90] Yang Xiaoti, Tang Qiao, Jiang Ying, et al. Nanoscale atp-responsive zeolitic imidazole framework-90 as a general platform for cytosolic protein delivery and genome editing. Journal of the American Chemical Society, 2019, 141: 3782-3786.

[91] Chen Guosheng, Huang Siming, Kou Xiaoxue, et al. A convenient and versatile amino-acid-aoosted biomimetic strategy for the nondestructive encapsulation of biomacromolecules within metal-organic frameworks. Angewandte Chemie International Edition, 2019, 58: 1463-1467.

[92] Lian Xizhen, Huang Yanyan, Zhu Yuanyuan, et al. Enzyme-MOF nanoreactor activates nontoxic paracetamol for cancer therapy. Angewandte Chemie International Edition, 2018, 57: 5725-5730.

[93] Oh J Y, Choi E, Jana B, et al. Protein-precoated surface of metal-organic framework nanoparticles for targeted delivery. Small, 2023, 19: 2300218.

[94] Sheng Chuangui, Yu Fangzhi, Feng Youming, et al. Near-infrared light triggered degradation of metal-organic frameworks for spatiotemporally-controlled protein release. Nano Today, 2023, 49: 101821.

[95] Liu Qian, Wang Li, Su Yitan, et al. Ultrahigh enzyme loading metal-organic frameworks for deep tissue pancreatic cancer photoimmunotherapy. Small, 2024, 20: 2305131.

[96] Katayama T, Tanaka S, Tsuruoka T, et al. Two-dimensional metal-organic framework-based cellular scaffolds with high protein adsorption, retention, and replenishment capabilities. ACS Applied Materials & Interfaces, 2022, 14: 34443-34454.

[97] Gong Peiwei, Cui Huiying, Li Cheng, et al. Self-stablized monodispersing nano-MOFs for controlled enzyme delivery. Chemical Engineering Journal, 2024, 489: 150941.

[98] Abuçafy M P, Frem R C, Polinario G, et al. MIL-100（Fe）sub-micrometric capsules as a dual drug delivery system. International Journal of Molecular Sciences, 2022, 23: 7670.

[99] Kim K, Lee S, Jin E J, et al. MOF × biopolymer: collaborative combination of metal-organic

framework and biopolymer for advanced anticancer therapy. ACS Applied Materials & Interfaces, 2019, 11: 27512-27520.

[100] Ha L, Choi K M, Kim D P. Interwoven MOF-coated janus cells as a novel carrier of toxic proteins. ACS Applied Materials & Interfaces, 2021, 13: 18545-18553.

[101] Yi Chengqiang, Zhu Lanxin, Li Dongyu, et al. Light field microscopy in biological imaging. Journal of Innovative Optical Health Sciences, 2022, 16: 2230017.

[102] Li Chenge, Tebo A G, Gautier A. Fluorogenic labeling strategies for biological imaging. International Journal of Molecular Sciences, 2017, 18: 1473.

[103] Allouche-Arnon H, Tirukoti N D, Bar-Shir A. MRI-based sensors for in vivo imaging of metal ions in biology. Israel Journal of Chemistry, 2017, 57: 843-853.

[104] Bianconi F, Palumbo I, Fravolini M L, et al. Form factors as potential imaging biomarkers to differentiate benign vs. malignant lung lesions on ct scans. Sensors, 2022, 22: 5044.

[105] Bieniek A, Terzyk A P, Wiśniewski M, et al. MOF materials as therapeutic agents, drug carriers, imaging agents and biosensors in cancer biomedicine: Recent advances and perspectives. Progress in Materials Science, 2021, 117: 100743.

[106] Zhao Lirong, Zhang Wei, Wu Qiong, et al. Lanthanide europium MOF nanocomposite as the theranostic nanoplatform for microwave thermo-chemotherapy and fluorescence imaging. Journal of Nanobiotechnology, 2022, 20: 133.

[107] Wagner P, Schwarzhaupt O, May M. In-situ X-ray computed tomography of composites subjected to fatigue loading. Materials Letters, 2019, 236: 128-130.

[108] Arora V, Sood A, Kumari S, et al. Hydrophobically modified sodium alginate conjugated plasmonic magnetic nanocomposites for drug delivery & magnetic resonance imaging. Materials Today Communications, 2020, 25: 101470.

[109] Ding Mengli, Liu Wenbo, Gref R. Nanoscale MOFs: From synthesis to drug delivery and theranostics applications. Advanced Drug Delivery Reviews, 2022, 190: 114496.

[110] Deng Jingjing, Wang Kai, Wang Ming, et al. Mitochondria targeted nanoscale zeolitic imidazole framework-90 for atp imaging in live cells. Journal of the American Chemical Society, 2017, 139: 5877-5882.

[111] Yang Chan, Chen Kun, Chen Mei, et al. Nanoscale metal-organic framework based two-photon sensing platform for bioimaging in live tissue. Analytical Chemistry, 2019, 91: 2727-2733.

[112] Robison L, Zhang Lin, Drout R J, et al. A bismuth metal-organic framework as a contrast agent for X-ray computed tomography. ACS Applied Bio Materials, 2019, 2: 1197-1203.

[113] Sene S, Marcos-Almaraz M T, Menguy N, et al. Maghemite-nanoMIL-100（Fe） bimodal nanovector as a platform for image-guided therapy. Chem, 2017, 3: 303-322.

[114] Wu Mingxue, Gao Jia, Wang Fang, et al. Multistimuli responsive core-shell nanoplatform constructed from Fe_3O_4@MOF equipped with pillar[6]arene nanovalves. Small, 2018, 14: 1704440.

[115] Chen Daiqin, Yang Dongzhi, Dougherty C A, et al. In vivo targeting and positron emission tomography imaging of tumor with intrinsically radioactive metal-organic frameworks nanomaterials. ACS Nano, 2017, 11: 4315-4327.

[116] Jiang Peichun, Hu Yang, Li Gongke. Biocompatible Au@Ag nanorod@ZIF-8 core-shell nanoparticles for surface-enhanced raman scattering imaging and drug delivery. Talanta, 2019, 200: 212-217.

[117] Li Yantao, Tang Jinglong, He Liangcan, et al. Core-shell upconversion nanoparticle@metal-organic framework nanoprobes for luminescent/magnetic dual-mode targeted imaging. Advanced Materials, 2015, 27: 4075-4080.

[118] Du Tianyu, Zhao Chunqiu, Rehman F, et al. In situ multimodality imaging of cancerous cells based on a selective performance of Fe^{2+}-adsorbed zeolitic imidazolate framework-8. Advanced Functional Materials, 2017, 27: 1603926.

[119] Shang Wenting, Zeng Chaoting, Du Yang, et al. Core-shell gold nanorod@metal-organic framework nanoprobes for multimodality diagnosis of glioma. Advanced Materials, 2017, 29: 1604381.

[120] Cai Wen, Gao Haiyan, Chu Chengchao, et al. Engineering phototheranostic nanoscale metal-organic frameworks for multimodal imaging-guided cancer therapy. ACS Applied Materials & Interfaces, 2017, 9: 2040-2051.

[121] Zhang Hui, Shang Yue, Li Yuhao, et al. Smart metal-organic framework-based nanoplatforms for imaging-guided precise chemotherapy. ACS Applied Materials & Interfaces, 2019, 11: 1886-1895.

[122] Xu Chen, Zhang Yukun, Sun Hui, et al. Development of a two-photon fluorescent probe for imaging hydrogen sulfide (H_2S) in living cells and zebrafish. Analytical Methods, 2023, 15: 1948-1952.

[123] Park K M, Kim H, Murray J, et al. A facile preparation method for nanosized MOFs as a multifunctional material for cellular imaging and drug delivery. Supramolecular Chemistry, 2017, 29: 441-445.

[124] Zhang Xiaolei, Li Sumei, Chen Sha, et al. Ammoniated MOF-74(Zn) derivatives as luminescent sensor for highly selective detection of tetrabromobisphenol a. Ecotoxicology and Environmental Safety, 2020, 187: 109821.

[125] Zhao Peiran, Liu Yuqian, He Cheng, et al. Synthesis of a lanthanide metal-organic framework and its fluorescent detection for phosphate group-based molecules such as adenosine

triphosphate. Inorganic Chemistry, 2022, 61: 3132-3140.

[126] Saikia K, Bhattacharya K, Sen D, et al. Solvent evaporation driven entrapment of magnetic nanoparticles in mesoporous frame for designing a highly efficient Mri Contrast Probe. Applied Surface Science, 2019, 464: 567-576.

[127] Wang Bo, Xu Fanfan, Zong Peijie, et al. Effects of heating rate on fast pyrolysis behavior and product distribution of jerusalem artichoke stalk by using Tg-ftir and Py-gc/Ms. Renewable Energy, 2019, 132: 486-496.

[128] Živanović M, Trenkić A A, Milošević V, et al. The role of magnetic resonance imaging in the diagnosis and prognosis of dementia. Biomolecules and Biomedicine, 2023, 23: 209-224.

[129] Lin Caixue, Sun Keke, Zhang Cheng, et al. Carbon dots embedded metal organic framework@chitosan core-shell nanoparticles for vitro dual mode imaging and pH-responsive drug delivery. Microporous and Mesoporous Materials, 2020, 293: 109775.

[130] Chen Meiling, Pang Shuchao, Chen Xuemin, et al. Synthesis of permeable yolk-shell structured gadolinium-doped quantum dots as a potential nanoscale multimodal-visible delivery system. Talanta, 2017, 175: 280-288.

第7章 MOFs未来发展与挑战

在过去的几十年里，MOFs 在其良好的结构、高表面积、高孔隙率、可调孔径和易于功能化等方面得到了广泛的研究。特别是，近年来，MOFs 作为纳米载体在生物医学中的应用引起了人们的极大兴趣。目前，各种分子已被研究作为疾病的治疗药物，如抗癌药物、核酸和蛋白质。虽然目前 NMOFs 作为药物载体的研究取得了一些进展，但其在临床医学中的实际应用仍有许多具有挑战性的工作有待完成。其可生物利用类型相对较少，并且对可以装载的药物类型有相对较大的限制。因此未来 NMOFs 和相关药物的类型可以扩大，使其适用于更多疾病的治疗。最后相关 MOFs 材料的质量可控规模化制备，也是目前限制其应用的重要瓶颈。

用于药物递送应用的多功能 NMOFs 工程是一项具有挑战性但很有前途的任务。需要不同学科的研究人员加强沟通与合作，为扩大 NMOFs 在生物医学领域的应用铺平道路，从而在促进人类疾病的预防、诊断和治疗研究方面发挥重要作用。如果上述问题能够在未来得到有效的解决，我们相信 NMOFs 将成为控制给药领域的核心载体，推进临床试验[1-3]。

监管问题是推进临床转化的关键步骤，任何临床 MOFs 必须首先符合监管机构（如 FDA）制定的严格的安全性和有效性标准。然而，临床试验是昂贵的，而且大家往往更愿意使用 FDA 批准的技术，而不是开发新的疗法。因此，转化思想对 MOFs 在生物医学中的未来至关重要。可喜的是基于 MOF 的癌症治疗系统已经开始进行临床试验（NCT0344444714）。该 MOF 给药系统由芝加哥的 Lin 团队开发，并由美国 RiMO 公司进一步研发，开发低剂量的 X 射线治疗和免疫治疗协同方案。

7.1 研究MOFs材料的生物安全性

研究药物代谢动力学主要是为了理解和预测药物在体内的行为，这对于药物的开发、优化、安全使用和治疗效果的评估至关重要。然而目前多数 MOFs 在药物递送领域的摄入后代谢机制尚不明确。

纳米药物临床应用的一个重要问题是生物安全。因此，基于 MOFs 系统的 NPs 毒性需要进行更系统的体内研究。确保生物安全的第一步是通过平衡无机和有机构件来建立 MOFs。无毒或低毒性的物质需要在体内和通过体内的代谢系统产生安全的分解产物。基于 MOFs 的控释药在给药后采用“ADME”机制，在生物体内进行的常用的小动物模型并不能预测药物的疗效和药物控释系统在人体中的性能。因此，必须建立创新的体内模型来准确地预测人体内部药代动力学的特征，这是成功向临床转化的巨大一步[3]。

基于 MOFs 的 DDS 临床应用的另一个主要挑战是其潜在的毒性。然而，现有的文献非常有限，无法得出关于 MOFs 纳米颗粒毒性的结论。到目前为止，已经对不同的细胞系进行了许多体外毒性研究，但是这些结果往往比较片面。例如，有研究用 nanoZIF-8（200 nm）对三种人类细胞系进行了评估，即 NCI-H292、HT-29 和 HL-60，结果表明，nanoZIF-8 对这些细胞无毒[4]。然而，在另一份报告中，nanoZIF-8（90 nm）显示出对 HeLa 和 J774 细胞系的细胞毒性[5]。最近，人们对纳米级 MOFs 的斑马鱼胚胎体内毒性进行了评估[6]。研究表明，MOFs 的毒性主要是由于浸出的金属离子。相反，三种不同的 Fe(Ⅲ)基 MOFs 纳米颗粒（MIL-88A、MIL-100 和 MIL-88B_4CH_3）被高剂量注射到大鼠体内。结果表明，这些 MOF 纳米颗粒具有低急性毒性，可被肝脾隔离。根据 Baati 等的研究，MOFs 纳米颗粒可以在尿液或粪便中进行进一步的生物降解和消除，而不发生代谢，引起显著毒性[7]。为了达到 MOF 纳米颗粒的临床开发阶段，需要对基于 MOF 的 DDS 进行系统的稳定性、降解力学和正常器官侧缺陷的体内研究，以进行临床前评价。

先进的 DDS 的出现为改善治疗方法的 ADME 特征开辟了新的视野。基于 MOFs 的 DDS 具有很有前景的特性，如增强载药能力、提高生物利用度和药物积累、提高半衰期、稳定给药浓度、减少给药频率、通过控制药物的部位特异性释放来减少药物在健康组织中的渗透，从而提高患者的依从性和便利性。例如多肽和抗体是有效的靶向部分，它们分别与活性受体和抗原具有较强的结合亲和力和更高的特异性相互作用[8]，虽然基于多肽-MOFs 的 DDS 报道很多，但是对其本身 ADME 报道却比较匮乏。

MOFs 的生物相容性一直是制约其广泛应用的关键因素之一。由于其在水溶液或生理环境中的稳定性相对较差，MOFs 容易出现迅速聚集或快速崩解的情况，这样的现象往往会引发细胞凋亡和组织异常等副作用。因此，如何在融合多种治疗方式的同时，显著提高 MOFs 的生物相容性，是当前研究者们需要着力解决的重要问题。只有实现了生物相容性的显著提升，MOFs 在生物医药领域的应用才能更加安全、有效。到目前为止，关于载药释放动力学的研究报道非常有限。载药过程受 MOFs 中笼的可及性控制，而装载能力受 MOFs 和药物分子的疏水性/亲水性影响[9]。

生物相容性金属有机框架（MOFs）由于其可调节的理化特性，已成为药物输送应用的潜在纳米载体。有报道具有可溶性金属中心的 Mg-MOF-74 可以促进某些药物的快速药代动力学。Pederneira 通过将不同量的布洛芬、5-氟尿嘧啶和姜黄素浸渍到 Mg-MOF-74 上[10]，研究了药物的溶解度如何影响药代动力学释放速率和递送效率。通过 HPLC 测试对 MOFs 在各种负载下的药物递送性能进行评估，结果表明释放速率是药物溶解度和分子大小的直接函数。在固定负载条件下考虑的三种药物中，负载 5-氟尿嘧啶的 MOFs 样品表现出最高的释放速率常数，这归因于 5-氟尿嘧啶相对于布洛芬和姜黄素具有最高的溶解度和最小的分子尺寸。值得注意的是，由于释放机制从单一化合物扩散模式转变为二元模式，释放动力学随着药物负载量的增加而降低。利用亲水配体的表面功能化可以提高 MOFs 的稳定性。

为了实现 MOFs 的临床潜力，未来正在进行的研究必须强调其稳定性、无毒性、生物相容性、技术可行性和有效性。在合成方面，研究者应强调其制备的绿色合成路线。此外，为了实现大规模的临床应用，还需要改进 MOFs 的合成方法。

7.2　合成方法的改进

尽管目前在 MOFs 合成领域取得了显著的成就，但仍有几个挑战有待解决。首先，虽然已经报道了许多功能化方法，但它们都有一些局限性。例如，表面吸附和孔隙包裹的分子由于相互作用力弱而逐渐泄漏。这些局限性需要开发先进的功能化策略，将多种潜在的治疗药物纳入 MOFs 中，以探索其临床应用。目前电化学法、微波法、机械化学方法、喷雾干燥和流动化学等都被报道用于制备 MOFs 材料（图 7.1），但是其大规模可控制备仍是难题[11-14]。

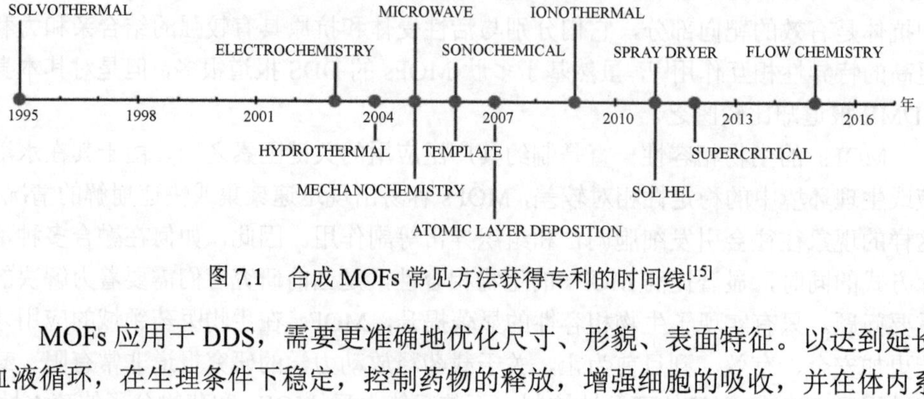

图 7.1　合成 MOFs 常见方法获得专利的时间线[15]

MOFs 应用于 DDS，需要更准确地优化尺寸、形貌、表面特征。以达到延长血液循环，在生理条件下稳定，控制药物的释放，增强细胞的吸收，并在体内系统中靶向给药的目的。许多类型的 MOFs 仍然局限于水溶液和缓冲溶液中，阻碍了药物控制释放。此外，MOFs 的合成主要是在高沸点的热条件下。它是在极性溶剂中进行的，如二甲基甲酰胺（DMF）和二乙基甲酰胺（DEF）。由于日益严重的环境问题和对可持续产品和工艺日益增长的需求，引入/开发更多的绿色方法（减少废物产生、有机无溶剂技术，产生更少的危险副产品）是一个关键[16]。

关于基于 MOFs 的超结构的表征，最近原位成像技术的使用，包括共聚焦显微镜和原位原子力显微镜（AFM），使自组装过程的实时可视化，从而探索自组装动力学。同样，原位小角度 X 射线散射（SAXS）提供了关于 MOFs 粒子空间排列的实时信息。因此，传统的技术如扫描电子显微镜（SEM）、透射电子显微镜（TEM）、扫描透射电子显微镜（STEM）、AFM 和 SAXS 都很好地补充了原位技术。

MOFs 基上层结构的实际开发受到其化学稳定性和对机械应力的敏感性的限制。基于 MOFs 的上层结构的稳定性尚未得到广泛的研究；因此，目前的数据集仍存在不足，同时，这些上层结构的低力学稳定性是不可否认的。研究重点在于束缚 MOFs 粒子并稳定空间结构的分子链，在相邻的 MOFs 粒子之间建立了相对较弱的相互作用。一般来说，这些弱相互作用对 MOFs 基超结构的力学性能有不利影响。因此，基于增强网状化学的自组装策略将确保上层结构的稳定性，因为这种理论方法是基于框架的强共价键。通过用高分子量和密度刷的聚合物接枝粒子，也可以提高上层结构的稳定性。Chin 等使用了类似的方法，他们通过光聚合 TMPTA 来固定嵌入在光聚合物中的定向 MOFs 粒子[17]。然而，到目前为止，这种方法可能会在增加机械稳定性和降低嵌入的 MOFs 粒子的上层结构顺序或固有孔隙度之间进行权衡。因此，需要进一步研究 MOFs 基上层结构的力学行为——包括硬度、弹性模量、剪切模量和屈服强度的测量——以全面了解影响其稳定性的变量。

要使 MOFs 真正应用于工业生产，批量合成无疑是关键的一步。为了实现这一目标，我们必须综合考虑以下多个因素：首先，起始原料必须便宜且容易获取，以确保生产成本的可控性；其次，合成条件应易于控制，理想情况下是在低温常压下进行，以减少能耗和设备要求；此外，反应步骤需要尽可能简单，以提高生产效率和降低成本；同时，活化过程也应简单易行，以缩短生产周期；当然，高产率和低杂质含量也是必不可少的，这关乎到产品的质量和市场竞争力；最后，所需溶剂量小，也是实现环保生产的重要一环。

近年来，MOF-5、Al-MIL-53、HKUST、ZIF-8 和 Fe-MIL-100 等材料已经成功实现了高通量的合成，并实现了商业化。例如，HKUST-1 的电化学批量化合成已经达到了超过 100 g 的量级，显示出巨大的生产潜力。而在微波辅助加热条件下，连续合成过程甚至可以达到 6.32×10^5 kg/(m^3·d)的产率[18]，这无疑是工业生产中的一个重大突破。这些成功的合成案例中，电化学、微波、水热、溶剂热等手段被广泛采用[19]，为 MOFs 的工业化生产提供了有力的技术支持。

控制 MOFs 粒子的自组装仍然是一项非常具有挑战性的任务。计算模型可能有助于控制 MOFs 粒子的自组装过程，指导实验合成工作，并最终设计新的超结构[8]。未来的工作应该着眼于扩大基于 MOFs 的超结构集。事实上，MOFs 粒子的自组装被限制在高度对称的结构中，如微球、线性链、周期晶格和周期性的三维结构。设计具有动态行为的上层结构的能力，能够与自然材料（如肌肉和细胞骨架）中发现的上层结构相媲美，这可能仍然是一个遥远的目标。同时，后续研究还可以集中于通过合成后的转化，包括煅烧过程，来扩展基于 MOFs 的超结构，以获得具有其他方面不可接近的性质的超结构。我们相信 MOFs 自组装领域将继续获得研究界的关注，并准备迎接增长和令人兴奋的进展。

7.3　拓展 MOFs 材料在药物递送领域的应用范围

拓展金属有机框架（MOFs）在药物递送领域的应用范围需要通过材料创新、技术集成以及跨学科合作来实现。以下是几种具体的策略，可用于推动 MOFs 在这一领域的进一步应用和发展[20]。

开发具有多种功能的 MOFs，如将靶向、成像和治疗功能集成到单一的 MOFs 结构中，实现一体化的疗法。通过精确控制 MOFs 的孔径、孔体积和表面功能化，优化药物的载荷量和释放动力学。复杂的肿瘤微环境（TME）和非特异性药物靶向限制了光动力学治疗联合化疗的临床疗效。据 Wang 等报道有金属有机框架（MOFs）辅助策略，可通过减少肿瘤缺氧和细胞内谷胱甘肽（GSH）来调节 TME，并提供靶向递送和受控释放的化学药物[21]。铂(Ⅳ)-二叠氮基络合物（Pt(Ⅳ)）装入基于羧酸铜(Ⅱ)的 MOF：MOF-199 中，在光照射下，释放的 Pt(Ⅳ)催化产生 O_2 可缓解缺氧并在癌细胞内产生基于 Pt(Ⅱ)的化学药物。同时，发射光敏剂 TBD 提供了有效的活性氧生成和明亮的发射，从而产生了以图像为导向的协同光化学疗法，具有增强的功效和减轻的副作用。Lin 和他的同事开发了 siRNA/UiO-Cis，用于联合传递顺铂和小干扰 RNA（siRNAs）治疗顺铂耐药卵巢癌[22]。利用 UiO 顺铂前药和 UiOFs 的高孔隙率和 Zr^{4+} 金属离子配位，进入 UiO 孔，将 siRNA 分别整合在 NMOFs 表面。目前 MOFs 材料在药物递送领域的应用已经从单一给药向着多功能方向发展。

推动学术界、医药企业和生物技术公司之间的合作，集中资源和专业知识来解决共同的科学和技术挑战。自 Nalco 化学公司 1995 年申请第一个专利以来，直到 2016 年由 Numat 技术公司发布了第一款基于 MOFs 的商业应用产品。MOFs 目前已实现大规模生产，这将有助于 MOFs 材料在药物递送应用领域的开发，并为其他基于 MOFs 的产品打开大门。不断增长的市场将推动 MOFs 供应商进一步提高成本效益、可再现性和环境可持续性，以保持竞争力（表 7.1）。例如，MOF Technologies 成立于 2013 年，是 UiO-66 和锆基 MOFs 家族的独家许可方。该公司专注于 MOFs 应用服务，旨在将研究和行业结合在一起，在气体存储、工业冷却、有毒气体保护和医疗保健等领域识别和开发商业上可行的应用。RiMO 公司成立于 2015 年 8 月，由芝加哥大学化学系和综合肿瘤中心林文斌博士创立，其新型“放疗–放射动态治疗（RT-RDT）”是全球首创的高效、低毒的革命性癌症治疗技术。目前国内亦有企业进入 MOFs 的生产应用领域。中科雷鸣可提供包括 ZIF-8（机械法水热法共沉淀法）、MIL-101、UiO-66、CuBTC 等多种 MOFs 材料。西安齐岳生物产品有 MOF-74、MIL-53-NH_2、UiO-66-F 等 MOFs 材料，不仅包括通用 MOFs 材料，而且可以通过-NH_2，-F 等进行进一步的功能化修饰。

表 7.1　MOFs 材料供应商

MOFs	供应商
Al（OH）fumarate	MOF Apps MOF Technologies
CAU-10	ProfMOF
Cu-BTC	BASF MOF Apps MOF Technologies
Fe-BTC	BASF
Magnesium formate	BASF MOF Technologies
Mg-MOF-74	MOF Technologies
MIL-100	KRICT MOF Apps
MIL-101-NH_2	MOF Apps
MIL-53	BASF
MIL-68	MOF Apps
MOF-177	BASF
PCN-250（Fe）	Framergy
UiO-66 series	Inven2 MOF Apps ProfMOF
ZIF-67	MOF Apps MOF Technologies
ZIF-8	BASF MOF Apps MOF Technologies STREM Chemicals Inc.
Zn-SIFSIX-pyrazine	MOF Technologies

用于生物医学应用的多功能 NMOFs 工程是一项具有挑战性但很有前途的任务。因此，不同学科的研究人员必须加强沟通与合作，为扩大 NMOF 在生物医学领域的应用铺平道路，从而在促进人类疾病的预防、诊断和治疗研究方面发挥重要作用。如果上述问题能够在未来得到有效的解决，我们相信 NMOFs 将成为控制给药领域的一种核心载体。

参 考 文 献

[1] Cao Jian, Li Xuejiao, Tian Hongqi. Metal-organic framework (MOF)-based drug delivery. Current Medicinal Chemistry, 2020, 27: 5949-5969.

[2] Moharramnejad M, Ehsani A, Shahi M, et al. MOF as nanoscale drug delivery devices: Synthesis and recent progress in biomedical applications. Journal of Drug Delivery Science and Technology, 2023, 81: 104285.

[3] Bigham A, Islami N, Khosravi A, et al. MOFs and MOF-based composites as next-generation. Materials for Wound Healing and Dressings Small, 2024, 20: 2311903.

[4] Bian Ruixin, Wang Tingting, Zhang Lingyu, et al. A combination of tri-modal cancer imaging and in vivo drug delivery by metal-organic framework based composite nanoparticles. Biomaterials Science, 2015, 3: 1270-1278.

[5] Wu Yafeng, Han Jianyu, Xue Peng, et al. Nano metal-organic framework (NMOF)-based strategies for multiplexed microrna detection in solution and living cancer cells. Nanoscale, 2015, 7: 1753-1759.

[6] Ruyra À, Yazdi A, Espín J, et al. Synthesis culture medium stability and *in vitro* and *in vivo* zebrafish embryo toxicity of metal-organic framework nanoparticles. Chemistry – A European Journal, 2015, 21: 2508-2518.

[7] Baati T, Njim L, Neffati F, et al. In depth analysis of the *in vivo* toxicity of nanoparticles of porous lron(III) metal-organic frameworks. Chemical Science, 2013, 4: 1597-1607.

[8] Rizvi S F A, Zhang Haixia, Fang Quan. Engineering peptide drug therapeutics through chemical conjugation and implication in clinics. Medicinal Research Reviews, 2024, 2024: 1-52.

[9] Rojas S, Colinet I, Cunha D, et al. Toward understanding drug incorporation and delivery from biocompatible metal-organic frameworks in view of cutaneous administration. ACS Omega, 2018, 3: 2994-3003.

[10] Pederneira N, Newport K, Lawson S, et al. Drug delivery on Mg-MOF-74: The effect of drug solubility on pharmacokinetics. ACS Applied Bio Materials, 2023, 6: 2477-2486.

[11] Munn A S, Dunne P W, Tang S V, et al. Lester, large-scale continuous hydrothermal production

and activation of ZIF-8. Chemistry Communication (Camb), 2015, 51: 12811-12814.

[12] Taddei M, Dau P V, Seth M, et al. Efficient microwave assisted synthesis of metal–organic framework UiO-66: Optimization and scale up. Dalton Transactions, 2015, 44: 14019-14026.

[13] Zhao Tian, Jeremias F, Ishtvan B, et al. High-yield, fluoride-free and large-scale synthesis of MIL-101(Cr). Dalton Transactions, 2015, 44: 16791-16801.

[14] Bayliss P A, Ibarra I A I, Perez E, et al. Synthesis of metal-organic frameworks by continuous flow. Green Chemistry, 2014, 6: 3796-3802.

[15] Rubio-Martinez M, Avci-Camur C, Thornton A W, et al. New synthetic routes towards MOF production at scale. Chemical Society Reviews, 2017, 46: 3453-3480.

[16] Batten M P, Rubio-Martinez M, Hadley T, et al. Continuous flow production of metal-organic frameworks. Current Opinion in Chemical Engineering, 2015, 8: 55-59.

[17] Yen S C, Lee Z H, Ni J S, et al. Effects of the number and position of methoxy substituents on triphenylamine-based chalcone visible-light-absorbing photoinitiators. Polymer Chemistry, 2022, 13: 3780-3789.

[18] Laybourn A, Lopez-Fernandez A M, Thomas-Hillman I, et al.Combining continuous flow oscillatory baffled reactors and microwave heating: Process intensification and accelerated synthesis of metal-organic frameworks. Chemical Engineering Journal, 2019, 356: 170-177.

[19] Rasmussen E G, Kramlich J, Igor V. Novosselov scalable continuous flow metal-organic framework (MOF) synthesis using supercritical CO_2. ACS Sustainable Chemistry & Engineering, 2020, 8: 9680-9689.

[20] Horcajada P, Chalati T, Serre C, et al. Porous metal-organic-framework nanoscale carriers as a potential platform for drug delivery and imaging. Nature Materials, 2010, 9: 172-178.

[21] Wang Yuanbo, Wu Wenbo, Mao Duo, et al. Metal-organic framework assisted and tumor microenvironment modulated synergistic image-guided photo-chemo therapy. Advanced Functional Materials, 2020, 30: 2002431.

[22] He Chunbai, Lu Kuangda, Liu Demin, et al. Nanoscale metal-organic frameworks for the CO-delivery of cisplatin and pooled sirnas to enhance therapeutic efficacy in drug-resistant ovarian cancer cells. Journal of the American Chemical Society, 2014, 136: 5181-5184.

编 后 记

“博士后文库”是汇集自然科学领域博士后研究人员优秀学术成果的系列丛书。“博士后文库”致力于打造专属于博士后学术创新的旗舰品牌，营造博士后百花齐放的学术氛围，提升博士后优秀成果的学术和社会影响力。

“博士后文库”出版资助工作开展以来，得到了全国博士后管委会办公室、中国博士后科学基金会、中国科学院、科学出版社等有关单位领导的大力支持，众多热心博士后事业的专家学者给予积极的建议，工作人员做了大量艰苦细致的工作。在此，我们一并表示感谢！

“博士后文库”编委会